ORYCHO
BASIC
RESEARCH

二月兰基础研究

张宝旭 主编

化学工业出版社
·北京·

内容简介

本书共4章，介绍了二月兰的形态、地理分布与种植、食用价值探讨、卫生学检验与毒理学评价、保肝作用研究等内容，旨在促进二月兰的药物及保健食品研发工作，为二月兰的相关生产实践提供帮助。本书采用中英文双语模式，并提供了大量二月兰相关照片，图文并茂。

本书可作为中药资源开发人员及经济植物研究和生产专业技术人员的参考书，也可供食品加工业从业者以及对保健食品感兴趣的读者参考阅读。

图书在版编目（CIP）数据

二月兰基础研究：英、汉/张宝旭主编. —北京：化学工业出版社，2021.7
ISBN 978-7-122-39596-2

I.①二… Ⅱ.①张… Ⅲ.①诸葛菜-研究-英、汉 Ⅳ.①S681.9

中国版本图书馆CIP数据核字（2021）第145239号

责任编辑：满孝涵　邱飞婵　　装帧设计：子鹏语衣
责任校对：李　爽　　封面设计：史利平

出版发行：化学工业出版社（北京市东城区青年湖南街13号　邮政编码100011）
印　　装：北京瑞禾彩色印刷有限公司
787mm×1092mm　1/16　印张12　字数350千字　2022年4月北京第1版第1次印刷

购书咨询：010-64518888　　售后服务：010-64518899
网　　址：http://www.cip.com.cn
凡购买本书，如有缺损质量问题，本社销售中心负责调换。

定　价：98.00元

编写人员名单

主编 张宝旭 北京大学公共卫生学院

编者 丁兆丰 北京汉华荣欣经贸有限公司

郭新慧 北京市房山区疾病预防控制中心

何月莹 北京药品监测中心

胡哲文 康龙化成（北京）生物技术有限公司

贾凤兰 北京大学公共卫生学院

敬 挺 深圳海关

李庆伟 辰晓投资（北京）管理有限公司

李荣佳 北京市东城区卫生健康监督所

李雪婷 国家卫生健康委老龄健康司

李园利 河南理工大学医学院

刘 庆 北京市海淀区疾病预防控制中心

刘伟霞 中国科学技术馆

刘晓晓 上海市青浦区疾病预防控制中心

刘 昕 深圳百年干细胞生物科技有限公司

吕 艳 中石化广州工程有限公司

吕颖坚 深圳市疾病预防控制中心营养与食品安全所

马秋霞 国家卫健委科研所

邱永祥 中国铁道科学研究院集团有限公司节能环保劳卫研究所

阮 明 北京大学公共卫生学院

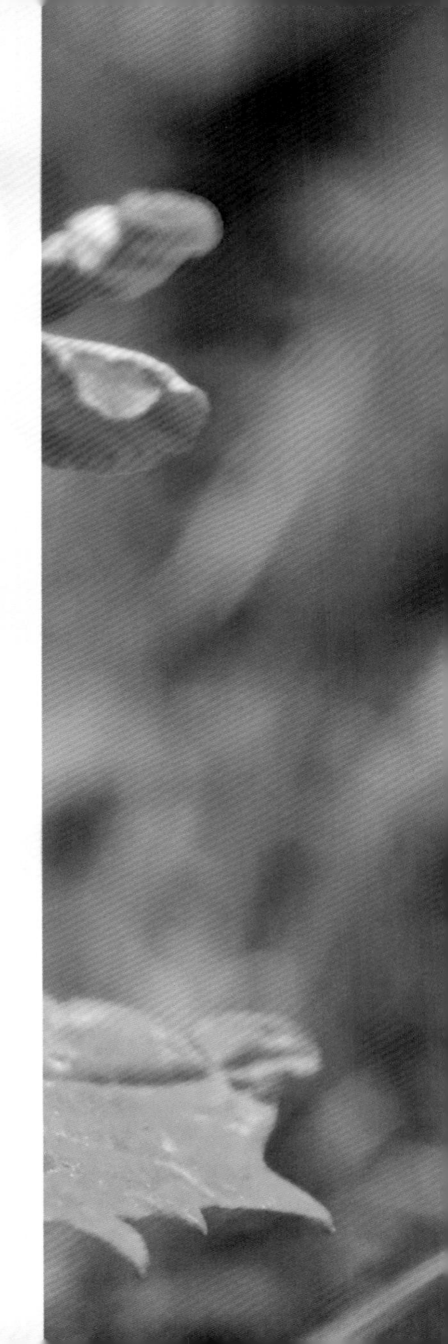

编写人员名单

沈金平　南通市疾病预防控制中心
宋　爽　中国疾病预防控制中心营养与健康所
万丽葵　中国疾病预防控制中心营养与健康所
王德伟　海关总署卫生检疫司
王福强　苏州药明康德新药开发有限公司
王　璐　北京海关
魏　鹏　上海费登斯投资咨询公司
魏雪涛　北京大学公共卫生学院
吴学银　成都苑东生物制药股份有限公司
邢国振　安领生物医药（深圳）有限公司
徐子茜　深圳市疾病预防控制中心
许旭东　中国医学科学院药用植物研究所
薛　茹　中国疾病预防控制中心辐射防护与核安全医学所
杨东旭　中邮保险河南分公司
战奕巍　北京市西城区月坛社区卫生服务中心
张宝旭　北京大学公共卫生学院
张梦萍　福建省疾病预防控制中心
朱乃亮　中国医学科学院药用植物研究所
祝靓靓　北京市海淀区卫健委

前言

二月兰是我们对数百种天然植物进行毒理学解毒剂研究时挖掘出来的保肝功效性植物。从初识二月兰到本书的编写，经历了十七年以上的系统性研究，在一定程度上揭开了二月兰的神秘面纱，可以让社会来共享我们的原创性科研成果，为人类卫生保健贡献我们的一份力量了。

本书的章节结构基本源于我们的实践，我们在校园进行二月兰种植实践、研究了二月兰的安全性和有用性以及有效成分单体。本书成书的过程也是培养研究生和广泛合作的过程。为了方便广大读者，本书采用了图文并举和中英文双语模式，尽量做到通俗易懂。

本书由国家重点研发计划“基于辨证保健的中药复方保健产品评价技术体系研究及示范平台的建立（2018YFC1706800）”资助。本书在编写过程中得到了全部作者的大力支持，也得到了出版社的热心帮助，在此一并感谢！

我们对二月兰的研究还在继续当中。由于水平有限，书中难免有不妥之处。诚恳期望读者不吝赐教，以便在本书再版时予以纠正。

张宝旭 医学博士
北京大学公共卫生学院毒理学系
北京大学国家中医药管理局中药配伍减毒重点研究室
Email：bxzhang@bjmu.edu.cn

Preface

Orycho (Orychophragmus Violaceus) is a plant with liver-protecting effect, which was excavated when we studied toxicological antidotes of hundreds of natural plants. After more than seventeen years' systematic research, the mystery of orycho has been unveiled to a certain extent, which allows the society to share our original scientific research achievements and contributes to human health promotion.

The organization of this book is basically derived from our practice. We planted orycho on campus, and studied the safety and usefulness of orycho and the monomer of the active ingredients. The process of completing this book is also a process of training graduate students and extensive cooperation. In order to facilitate the readers, this book is well-illustrated, bilingual in both Chinese and English, so as to make it easy to understand.

The book is funded by the National Key R&D Program (2018YFC1706800). In the process of writing this book, we received strong support from all the authors, and kind assistance from the publishing house. Thanks here!

Our research on orycho is still ongoing. Due to the limited level, mistakes are inevitable in the book. I sincerely hope that readers will kindly give us advice so as to improve this book when it is reprinted.

Baoxu Zhang, M,D.
Department of Toxicology, School of Public Health, Peking University
Peking university state administration of traditional chinese medicine primary laboratory of Chinese medicine compatibility and detoxification
Email: bxzhang@bjmu.edu.cn

目　录　CONTENTS

第一章

第二章

目 录 CONTENTS

第三章

第四章

第一章 二月兰概述 Overview of orycho

二月兰的名称虽然好听，但是并不是现代植物学分类中的兰科（Orchid）的植物（图 1-1），而是古老民间对该花卉的一种称呼。在本书中有时叫二月兰，有时叫诸葛菜，仅仅是为了表述的方便而已。

通用名：二月兰。

学名：诸葛菜（中国植物志）。

俗名：菲葸菜、二月蓝。

拉丁学名：*Orychophragmus Violaceus*。

英文名：Orycho（我们利用了拉丁名的前六个字母来使用的外文名字，发音近似二月兰），Chinese violet cress。

花语：仁爱、热情、智慧源泉、优秀。

Although the Chinese name of eryuelan (February orchid), is pleasant to hear, it is not an Orchidaceae plant in modern botanical classification (Fig.1-1), but an ancient folk name. In this book, it is sometimes called orycho or zhugecai, just for the convenience of expression.

Common name in Chinese: Eryuelan.

Scientific name in Chinese: Zhugecai (Flora of China).

Local name in Chinese: Feixicai, eryuelan (February royal purple).

Latin scientific name: *Orycho phragmus Violaceus*.

English name: Orycho (we made), Chinese violet cress.

Flower Language: Humanity, Enthusiasm, Fountain of Wisdom, Excellence.

图 1-1 兰花（六片花瓣）
Fig.1-1 Orchid（six petals）

二月兰的形态

北京的二月兰为跨年的二年生草本，其规律基本是秋苗、冬苞 、春花 、夏实（图 1-2）。苗期为基生叶，叶近圆形或短卵形。花期为抱茎 叶窄卵形为主（图 1-3）。叶边缘有钝齿。花瓣 4 片，分离，成十字形排列，花瓣有白色、粉红色、淡紫色、淡紫红色或紫色等，紫色为 主。雄蕊 6 个，雌蕊 1 个，均为黄色。

Morphology of orycho

Orycho in Beijing is a biennial herb, and its life cycle laws are basically autumn seedlings, winter buds, spring flowers and summer fruits (Fig. 1-2). At seedling stage, the leaves are basal leaves, which are nearly round or short oval. The flowering period is mainly narrow oval of clasped stems and leaves (Fig.1-3). The leaf edge has blunt teeth. There are 4 petals, which are separated and arranged in a cross shape. The petals are white, pink, lavender, mauve or purple, and purple is the main one. There are 6 stamens and 1 pistil, all of which are yellow.

图 1-2 北京地区二月兰的一生
Fig.1-2 Life of orycho in Beijing

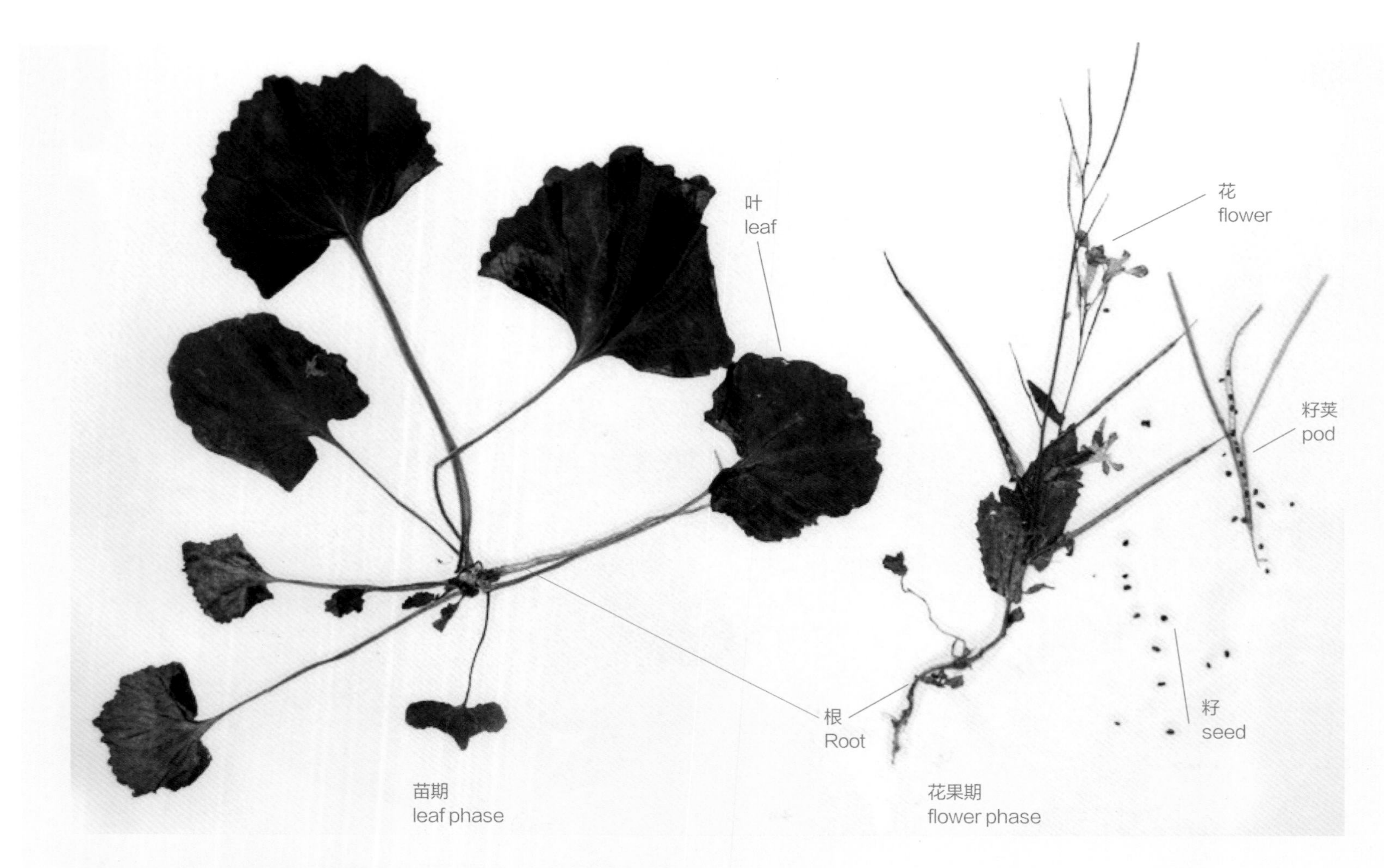

图 1-3 跨年（二年）生二月兰的标本，叶子在秋天（左）和春天开花时（右）
Fig.1-3 A specimen of orycho life (biennial), whose leaves bloom in autumn (left) and spring (right)

二月兰的地理分布

我们在 2019 年进行的统计显示，中国除了西藏、 海南和台湾以外，都是二月兰的产地。各地都可以发展二月兰种植产业。

Geographical distribution of orycho

Our 2019 statistics show that China is the origin of orycho except Tibet, Hainan and Taiwan Province. All regions can develop the planting industry of orycho.

二月兰的种植

二月兰田间种植

菜地的选择

选地原则：首选基本农田，建立种植日记制度。

搞清环境背景及可能的污染来源。最好选择经过“蔬菜生产基地环境质量监测”的地块，预防有害金属和非金属等污染物，要遵守优质农业操作规范（Good Agriculture Practice, GAP）（图 1-4 和图 1-5）。

Planting of orycho

Planting orycho in field

Choice of fields

The principle of land selection: The first choice is basic farmland, and the planting diary system is established.

Find out the environmental background and possible pollution sources. It is best to select plots that have passed the environmental quality monitoring of vegetable production bases to prevent harmful metal and nonmetal pollutants, and to comply with Good Agriculture Practice (GAP) (Fig.1-4 and Fig.1-5).

图 1-4 二月兰种植选择基本农田
Fig.1-4 Choice basic farmland

图 1-5 选择有野鸟的农田
Fig.1-5 Choose farmland with wild birds

播种和田间管理

农业投入品使用原则：原则上不使用。如果必须使用时，一定在农业技术站专家指导下，制定严格的无公害使用标准后，按规定使用农药，严格控制农药残留，确保生产无公害二月兰产品。

播种和生长过程要精细管理（图 1-6~ 图 1-8）。

收获

适时收获二月兰叶菜和菜薹。

叶菜收获标准：当叶和菜薹生长到 10 厘米以上时开始收获（摘取），一直可以收获到菜薹期。按照北京市标准《蔬菜采后处理技术规程第 2 部分：叶菜类（DB11/ T 867.2—2012）》进行（图 1-9~ 图 1-13）。

每亩二月兰的产量可以达到一千公斤以上。

图 1-6 二月兰机械化播种
Fig.1-6 Mechanized sowing

图 1-7 及时人工浇水
Fig.1-7 Timely artificial watering

图 1-8 人工除草
Fig.1-8 Artificial weeding

图 1-9 进入收获期（第 35 天），菜叶长到 10 厘米
Fig.1-9 At harvest time (the 35th day), the leaves grow to 10 cm

Sowing and field management

Use principle of agricultural inputs: in principle, they are not used. If it must be used, under the guidance of agricultural technology station experts, strict pollution-free use standards must be formulated, pesticides should be used according to regulations, and pesticide residues should be strictly controlled to ensure the production of pollution-free orycho products.

Carefully sowing and growth process management (Fig.1-6~Fig.1-8).

Harvest

Harvest orycho leaves and flower stalks in due course.

Harvesting standard of leafy vegetables: When leaves and flower stalks grow to more than 10 cm, they will be harvested (picked), and can be harvested until the flower stalks stage. According to Beijing standard "Technological standards for postharvest handling of vegetables Part 2: Leafy vegetables (DB11/T 867.2-2012)" (Fig.1-9~Fig.1-13). The yield of orycho per mu can reach more than 1 000kg.

图 1-10 细心收获
Fig.1-10 Careful harvest

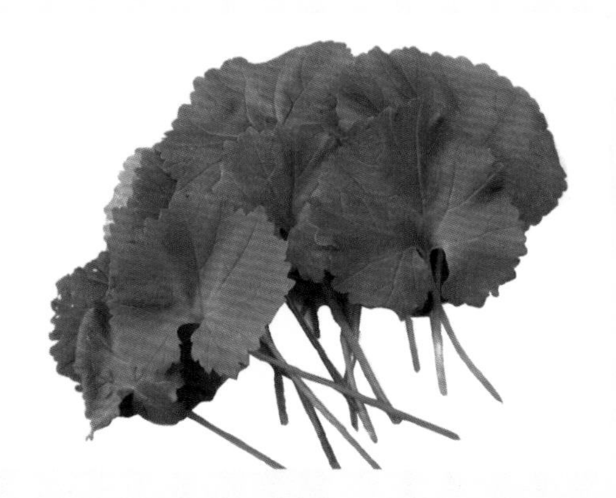

图 1-11 二月兰嫩叶
Fig.1-11 Young leaves

图 1-12 菜薹期
Fig.1-12 Tender flower stalk

图 1-13 二月兰鲜花
Fig.1-13 Flowering stage

二月兰林下种植

二月兰林下种植是实现立体农业的一种好方式，可以充分利用农业空间，增加收入。

Planting orycho under forest

Planting orycho under forest is a good way to realize three-dimensional agriculture, which can make full use of agricultural space and increase income.

二月兰水培

要保障二月兰的不间断供应，必须发展二月兰的植物工厂技术。我们进行了二月兰水培试验，取得了初步的成功。下一步，将试验人工照明和大型栽培床等试验。

Hydroponics of orycho

To ensure the uninterrupted supply of orycho, it is necessary to develop the plant factory technology of orycho. We carried out the experiment of hydroponics of orycho and achieved initial success. Next, we will test artificial lighting and large-scale cultivation bed.

二月兰资源相关农业技术发展前景

二月兰开花持续时间长，花朵多，又是林下经济，可以将养蜂场建在二月兰基地来获取蜂蜜，并利用蜜蜂来授粉。二月兰蜂蜜的成分和价值还需要评估。应该发展二月兰蜂蜜产业，以期生产出一种独特新品种蜂蜜。

Prospect of agricultural activity related to orycho resources

Orycho blooms for a long time with many flowers, which is also an under- forest economy. Beekeeping farms can be set in orycho base to obtain honey and use bees for pollination. The composition and value of orycho honey still need to be evaluated. The orycho honey industry should be developed in that hope of producing a unique new variety of honey.

第二章 二月兰食用价值 Edible value of orycho

二月兰的食用历史——被遗忘的保健蔬菜

二月兰历史悠久（图2-1）。我们从历史文献中检索到的二月兰植物绘图记录中，有173年前（1848年）清朝时期出版的《植物名实图考》（图2-2），此时已经将二月兰和蔓菁（芜菁）完全分开为各自独立的植物。

Edible history of orycho——a forgotten health vegetable

Orycho has a long history (Fig. 2-1). Among the plant drawings we have retrieved from the historical literature are the "Illustrated Investigation of the Names and Natures of Plants (ZhiwuMing Shi Tukao)" published 173 years ago (1848) during the Qing Dynasty (Fig. 2-2), in which the orycho and turnip have been completely separated into different species.

历史年代 Era	历史记载 Historical records
（公元 220 年—280 年）三国时期 (AD 220—280) Three Kingdoms Period	诸葛亮时代（批准推广作为军食） Zhuge Liang era (approved for promotion as military food)
（公元 856 年）唐朝《刘公嘉话录》 诸葛菜和蔓菁（芜菁）混为同种 (AD 856)" Liu Gong Jia Hua Lu "of Tang Dynasty Zhugecai and turnip are mixed into the same species	记载了诸葛亮在蜀汉国时的传说 芽、叶、根、籽都可食（现在同） It records the legend of Zhuge Liang in Shu Han Buds, leaves, roots and seeds were all edible (now the same)
（公元 1848 年）清朝《植物名实图考》诸葛菜与蔓菁种属分开，进入现代植物学分类阶段，各自独立 (AD 1848) In the Qing Dynasty, "The Textual Research on the Names and Facts of Plants", genera of zhugecai and turnip were separated and entered the stage of modern botanical classification, and they were independent of each other	作者吴其濬，插图和现代诸葛菜一样 诸葛菜和蔓菁叶逐步沦为野菜 Author Wu Qijun, the illustration is the same as modern The leaves of orycho and turnip gradually become wild vegetables
（公元 1955 年）《经济植物手册》新中国时代 (AD 1955) ,"Handbook of Economic Plants", in New China Era	胡先骕编，名称有菲葸菜、诸葛菜、二月蓝 Edited by Hu Xiansu, the names are Feixicai, Zhugecai and orycho (February royal purple)
（公元 1987 年）《中国植物志》现代 (AD 1987)，" Flora of China"，modern	国家层次上正式命名为诸葛菜 Officially named Zhugecai at the national level

图 2-1 二月兰（诸葛菜）食用历史进程
Fig.2-1 Brief edible historical process of orycho

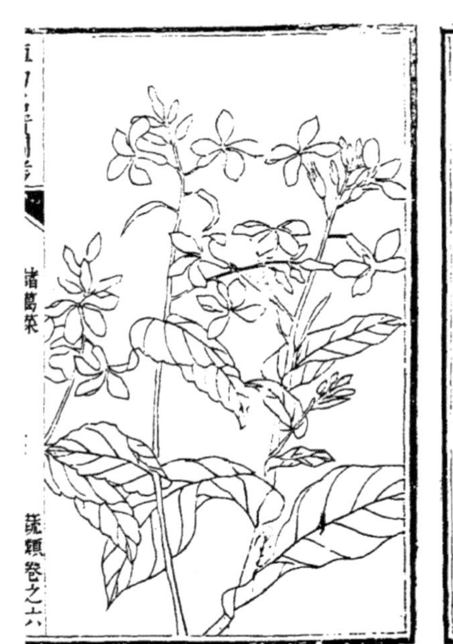

植物名實圖考

諸葛菜

諸葛菜北地極多湖南間有之初生葉如小葵抽葶生葉如油菜莖上葉微寬有圓齒亦抱莖生春初開四瓣紫花頗嬌亦有白花者耐霜喜寒京師二月已舒葶矣瀹食甚滑細根非蔓菁一名諸葛菜也按爾雅菲蒠菜郭注菲草生下濕地似蕪菁華紫赤色可食陸璣詩疏菲似葍莖麤葉厚而長有毛三月中蒸鬻為茹滑美可作羹幽州人謂之芴今河內人謂之宿菜按其形狀正是此菜北地至多皆生廢圃中無種植者因宿根而生故呼宿菜不知何時誤呼諸葛也江西有一種藤菜與此相類

图 2-2 清朝《植物名实图考》（1848 年出版）
Fig.2-2 "Textual Research on the Names and Facts of Plants" in Qing Dynasty (published in 1848)

1953 年胡先骕编写的《经济植物手册》，指出二月兰“味甚美”（图 2-3）。

"Handbook of Economic Plants", written by Hu Xiansu in 1953. It is pointed out that orycho is "very delicious vegetable" (Fig.2-3).

25.菲菜菜 Orychophragmus violaceus, O. E. Schultz.

此種爲無毛一年生草本植物；莖細瘦，分枝甚繁，高30—60公分。葉無柄，基部有葉耳；根生葉存在不久，羽狀分裂，頂端裂片大，心臟形，側生裂片四五對，頗小，有波狀齒；莖生葉倒卵矩圓形，頂端微尖，有波狀鋸齒；花直徑30公分，淡紫色，成疏鬆頂生花序；花梗細瘦，無苞片；萼片長18公厘，直立，兩側兩個基部成囊狀，與花瓣之爪等長；花瓣片圓倒卵形，全緣，鋪張；角細瘦，長 7.5—8 公分，鈍四邊形；裂瓣有突起中脈；喙短圓錐狀；種子成一行，小，矩圓形，黑色。

此種產於亞洲，在我國自東北至長江以南皆產；可作蔬菜用，味甚美，又名宿菜，北京名爲二月蘭，南京名爲諸葛菜。

图 2-3 《经济植物手册》（1955 年出版）

Fig.2-3 "Handbook of Economic Plants" published in 1955

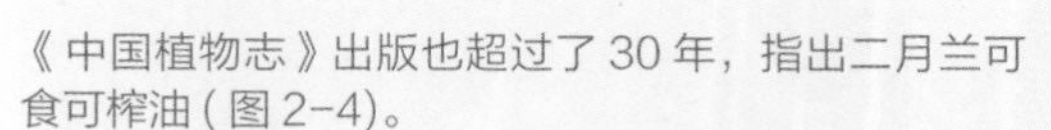

《中国植物志》出版也超过了 30 年，指出二月兰可食可榨油（图 2-4）。

"Flora of China" has been published for more than 30 years, pointing out that orycho is edible and is used to extract oil (Fig.2-4).

1a. 诸葛菜(原变种) 图版 7:5—7

Orychophragmus violaceus (L.) O. E. Schulz var **violaceus**

一年或二年生草本，高 10—50 厘米，无毛；茎单一，直立，基部或上部稍有分枝，浅绿色或带紫色。基生叶及下部茎生叶大头羽状全裂，顶裂片近圆形或短卵形，长 3—7 厘米，宽 2—3.5 厘米，顶端钝，基部心形，有钝齿，侧裂片 2—6 对，卵形或三角状卵形，长 3—10 毫米，越向下越小，偶在叶轴上杂有极小裂片，全缘或有牙齿，叶柄长 2—4 厘米，疏生细柔毛；上部叶长圆形或窄卵形，长 4—9 厘米，顶端急尖，基部耳状，抱茎，边缘有不整齐牙齿。花紫色、浅红色或褪成白色，直径 2—4 厘米；花梗长 5—10 毫米；花萼筒状，紫色，萼片长约 3 毫米；花瓣宽倒卵形，长 1—1.5 厘米，宽 7—15 毫米，密生细脉纹，爪长 3—6 毫米。长角果线形，长 7—10 厘米，具 4 棱，裂瓣有 1 凸出中脊，喙长 1.5—2.5 厘米；果梗长 8—15 毫米。种子卵形至长圆形，长约 2 毫米，稍扁平，黑棕色，有纵条纹。 花期 4—5 月，果期 5—6 月。

产辽宁、河北、山西、山东、河南、安徽、江苏、浙江、湖北、江西、陕西、甘肃、四川。生在平原、山地、路旁或地边。朝鲜有分布。模式标本采自中国。

嫩茎叶用开水泡后，再放在冷开水中浸泡，直至无苦味时即可炒食。种子可榨油。

图 2-4 《 中国植物志 》(1987 年出版）

Fig.2-4 "Flora of China" published in 1987

中国数字植物标本馆（http://www.cvh.ac.cn/）中可以查到很多二月兰食用的植物标本（图 2-5~ 图 2-6）。

A large number of plant specimens from chinese virtual herbarium (http://www.cvh.ac.cn/) have been recorded for the consumption of orycho(Fig.2-5~Fig.2-6).

图 2-5 北京大学二月兰植物标本 (1921 年）
Fig.2-5 Plant specimen of orycho at Peking University (1921)

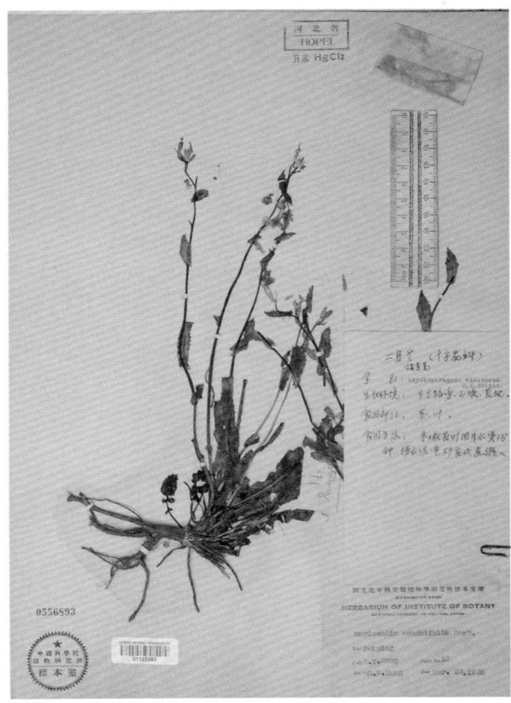

图 2-6 1935 年记录二月兰食用方法（炒食煮粥）的植物标本
Fig.2-6 Plant specimen of orycho recording edible method in 1935

二月兰作为食用农产品（野菜）的现况

农业部网站明确二月兰种植可为市民提供新鲜野菜。

2014 年 7 月 18 日，农业部网站上指出在玉米地中种植二月兰不仅可以作为绿肥而且为市民提供了早春野菜。二月兰鲜嫩菜薹含有丰富的胡萝卜素和维生素等营养成分。可以肯定地说，作为蔬菜种植生产的国家主管部委，农业部认可二月兰是蔬菜，是一种可食用农产品。

Current situation of orycho as edible agricultural product(wild vegetables)

The website of the Ministry of Agriculture specifies that orycho planting can provide fresh wild vegetables for citizens.

On July 18, 2014, the website of the Ministry of Agriculture pointed out that planting orycho in corn fields can not only be used as green fertilizer but also provide citizens with wild vegetables in early spring. The fresh and tender flower stalk is rich in nutrition such as carotene and vitamins. To be sure, as the national competent department in charge of vegetable planting and production, the Ministry of Agriculture and Rural Affairs of the People's Republic of China has recognized that orycho is a vegetable and an edible agricultural product.

您现在的位置:首页>北京

北京市种植二月兰绿肥作物效果显著

日期：2014-07-18 13:25 作者： 来源：北京市农业局 点击次数： 90

北京市在京郊果园及春玉米种植区域累计种植二月兰绿肥作物30多万亩，效果显著：一是提高土壤有机质含量，改善耕地质量。二月兰每亩鲜草产量1500多公斤，连续翻压五年提高土壤有机质含量20%，每亩补充土壤氮磷钾纯养分12.8公斤左右，土壤容重平均降低10.8%，土壤保水、保肥能力增强，缓冲能力提高。二是提高农作物及果品产量，改善了果品品质，提高了化肥利用效率。春玉米二月兰复种、果园二月兰间作两种模式分别增产11.2%、7.5%以上，替代化肥用量16.0-22.0%、13.0-15%，提高肥料利用率10.0-20.2%、15.0-22.5%，果品VC含量提高23.5%。三是为创造优美田间景观增加新内容。四是覆盖裸露土壤，抑制风沙扬尘，改善大气环境。二月兰秋季播种后，整个冬季覆盖率可以达到80%，4月底覆盖率可以达到100%，有效覆盖裸露土壤。五是为市民提供了早春野菜。鲜嫩菜薹含有丰富的胡萝卜素和维生素等营养。

会讯公告

关于举办科
关于举办科
关于举办调
关于举办农
北京润生农
中华人民共
2017年度农
关于对拟表
关于对拟认

招标公告

二月兰是春天市民争先采摘的菜蔬。虽然现在市场上没有二月兰出售，但是每到春天市民都会到郊区采摘二月兰，特别是在北京，市民到公园采摘二月兰等野菜成为媒体报道的热点。公园不得不张贴禁止采摘二月兰的告示。

Orycho is the vegetable that the citizens rush to pick in spring. Although there is no orycho for sale on the market, every spring, citizens will go to the suburbs to harvest orycho. Especially in Beijing, citizens go to parks to harvest orycho and other wild vegetables, which has become a hot topic in media reports. The park had to put up notices forbidding the picking of orycho.

北京教学植物园的二月兰标牌告诉人们它的嫩叶可以食用，种子可以榨油。

Orycho sign in Beijing Teaching Botanical Garden tells people that its fresh leaf is edible, its seeds can be used to made cooking oil.

二月兰的营养成分

二月兰叶和籽的成分

二月兰叶和籽含有全面的五大营养素（蛋白质、脂肪、碳水化合物、维生素、矿物质）。叶含有丰富的蛋白质，籽含有丰富的脂肪酸（表2-1~表2-5）。

Nutrient composition of orycho

Composition of orycho leaf and seed

Orycho leaves and seeds contain five comprehensive nutrients (protein, fat, carbohydrate, vitamins and minerals). Leaves are rich in protein and seeds are rich in fatty acids (Table 2-1~Table 2-5).

表 2-1 北京二月兰叶粉和籽的营养成分表
Table 2-1 Composition of Orycho Leaf Powder and Seed

成分 Composition	单位 Unit	含量 Content	
		干叶粉 Leaf powder	籽 Seed
可食部 Edible	%	100	100
能量 Energy/kcal	kcal/110g	405	763
能量 Energy/kJ	kJ/110g	1694	3193
碳水化合物 Carbohydrates	g/100g	38.59	24.74
蛋白质 Protein	g/100g	34.5	22.5
脂肪 Fat	g/100g	5.6	44.4
灰分 Ash	g/100g	14.6	3.4
水分 Water	g/100g	6.71	4.96
总膳食纤维 Dietary fiber	g/100g	27.9	19.4
δ-生育酚 δ-tocopherol	mg/100g	0.999	0.274
γ-生育酚 γ-tocopherol	mg/100g	4.01	1.8
α-生育酚 α-tocopherol	mg/100g	24.0	27.0
维生素 E Vitamin E	mg/100g	29.0	30.80
维生素 K_1 Vitamin K_1	mg/100g	1.19	62.50
β-胡萝卜素 β-carotene	mg/100g	7.33	47.60
维生素 C Vitamin C	mg/100g	148	<0.004
维生素 B_1 Vitamin B_1	mg/100g	< 0.10	0.142
维生素 B_2 Vitamin B_2	mg/100g	0.735	0.090
维生素 B_6 Vitamin B_6	mg/100g	0.88	0.72
烟酸 Niacin	mg/100g	5.8	7.6
叶酸 Folate	mg/100g	1.37	0.15

表 2-2 北京二月兰叶粉和籽的氨基酸
Table 2-2 Amino acids in orycho leaf powder and seed

氨基酸 Amino Acid	单位 Unit	含量 Content	
		干叶粉 Leaf powder	籽 Seed
天门冬氨酸 ASP	g/100g	2.82	1.34
苏氨酸 THR	g/100g	1.29	0.42
丝氨酸 SER	g/100g	1.39	0.71
谷氨酸 GLU	g/100g	6.33	2.84
甘氨酸 GLY	g/100g	1.30	1.17
丙氨酸 ALA	g/100g	1.37	0.66
缬氨酸 VAL	g/100g	1.76	0.31
蛋氨酸 MET	g/100g	0.36	0.26
异亮氨酸 ILE	g/100g	1.15	0.17
亮氨酸 LEU	g/100g	2.07	0.80
酪氨酸 TYR	g/100g	1.03	0.59
苯丙氨酸 PHE	g/100g	1.50	0.78
赖氨酸 LYS	g/100g	1.65	0.65
组氨酸 HIS	g/100g	0.76	0.28
精氨酸 ARG	g/100g	1.49	0.90
脯氨酸 PRO	g/100g	3.95	0.95
色氨酸 TRY	g/100g	0.43	0.31
胱氨酸 CYS	g/100g	0.43	0.41

表 2-3 北京二月兰叶粉和籽食物成分表（矿物质）
Table 2-3 Composition of orycho leaf powder and seed (minerals)

成分 Composition	单位 Unit	含量 Content	
		干叶粉 Leaf powder	籽 Seed
磷 Phosphorus	mg/100g	612	3870
钙 Calcium	g/kg	28.5	2.8
铜 Copper	mg/kg	4.25	2.98
铁 Iron	mg/kg	200	128
钾 Potassium	g/kg	37.7	9.1
镁 Magnesium	g/kg	2.78	2.56
锰 Manganese	mg/kg	42.8	20
钠 Sodium	mg/kg	531	24.1
锌 Zinc	mg/kg	36.9	36.3
硒 Selenium	mg/kg	0.054	0.079

表 2-4 二月兰籽油脂肪酸组成
Table 2-4 Fatty acid composition of orycho seed oil

序号	脂肪酸	Fatty Acid	含量 Content (mg/g)	百分比 Percentage (%)
1	棕榈酸 (16 ∶ 0)	Palmitic Acid (16 ∶ 0)	91.53±0.59	21.7
2	硬脂酸 (18 ∶ 0)	Stearic Acid (18 ∶ 0)	70.71±1.07	16.7
3	油酸 (18 ∶ 1*n* − 9)	Oleic Acid (18 ∶ 1*n* − 9)	70.46±0.51	16.7
4	异油酸 (18 ∶ 1*n* − 10)	Isooleic Acid (18 ∶ 1*n* − 10)	8.94±0.92	2.1
5	亚油酸 (18 ∶ 2)	Linoleic Acid (18 ∶ 2)	103.72±1.02	24.6
6	花生酸 (20 ∶ 0)	Arachidic Acid (20 ∶ 0)	9.10±0.05	2.2
7	顺 -11- 二十碳烯酸 (20 ∶ 1)	Cis-11-Eicosenoic Acid (20 ∶ 1)	12.96±0.24	3.1
8	亚麻酸 (18 ∶ 3*n* − 3)	Linolenic Acid (18 ∶ 3*n* − 3)	24.01±0.58	5.7
9	顺 -11,14- 二十碳烯酸 (20 ∶ 2)	Cis-11,14-Eicosadienoic Acid (20 ∶ 2)	2.09±0.00	0.5
10	二十二烷酸 (22 ∶ 0)	Behenic Acid (22 ∶ 0)	6.53±0.21	1.5
11	芥酸 (22 ∶ 1)	Erucic Acid (22 ∶ 1)	1.35±0.05	0.3
12	木蜡酸 (24 ∶ 0)	Lignoceric Acid(24 ∶ 0)	1.76±0.01	0.4

表 2-5 二月兰籽油风味挥发性成分化学组成
Table 2-5 Composition of flavor volatile components in orycho seed oil

序号	化合物	Compound
1	庚醛	Heptanal
2	正己醛	Hexanal
3	2- 己醛	2-Hexenal
4	2- 庚酮	2-Heptanone
5	3- 蒈烯	3-Carene
6	α - 蒎烯	α -Pinene
7	莰烯	Camphene
8	蒎烯	Bicyclo Heptane
9	己酸	Hexanoic acid
10	2,4- 己二醛	2,4-Heptandienal
11	2,4- 己二烯	2,4-Hexadiene
12	邻伞花烃	o-Cymene
13	D- 苧烯	D-limonene
14	3,5- 辛二烯 -2- 醇	3,5-Octadien-2-ol
15	2- 辛烯醛	2-Octenal
16	3,5- 辛二烯 -2- 酮	3,5-Octadien-2-one
17	壬醛	Nonanal
18	苯乙醇	Phenylethyl Alcohol
19	1,2- 二甲氧基苯	Benzene,1,2-dimethoxy
20	2- 壬烯醛	2-Nonenal
21	2- 甲氧基 -3- (1- 甲基异丙基) 吡嗪	2-methoxy-3-(1- methylpropyl)-Pyrazine
22	萘	Naphthalene
23	2,4- 壬二烯醛	2,4-Nonadienal
24	2- 癸烯醛	2-Decenal
25	2,4- 癸二烯醛	2,4-Decadienal
26	十三烷	Tridecane
27	环癸酮	Cyclodecanone
28	长叶蒎烯	Tricyclo[5.4.0.0(2,8)] undec-9-ene,2,6,6,9-tetramethyl
29	长叶松烯	Longifolene-(V4)
30	α - 长蒎烯	α -Longipinene
31	长叶烯	Longifolene
32	顺 - 香柠檬烯	Cis-Bergamotene
33	顺 - 罗汉柏烯	Cis-Thujopsene
34	雪松烯	Cedrene
35	香橙烯	Aromandendrene
36	依兰烯	α -ylangene
37	7- 表 - 顺 - 水合倍半香桧烯	7-epi-cis-sesquisabinene hydrate

二月兰籽苦味物质等的探讨

二月兰叶和二月兰籽都有一定的苦味。我们对二月兰籽的水溶性成分进行检测，一共检出了水杨酸等微量的苦味的植物成分，包括布卢门醇、3',5',5,7- 四羟基二氢黄酮、儿茶素、槲皮素、腺苷、水杨酸 、2' - 脱氧腺苷、对羟基桂皮酸、香草酸、原儿茶酸、咖啡酸、丁香酸、胡萝卜苷、丁二酸和壬二酸。这些植物成分对二月兰风味的贡献需要进一步研究。

Composition of bitter substances in orycho seed

Orycho leaf and seed have a certain bitter taste. The water-soluble components of the seeds were detected, and some trace plant components with bitter taste such as salicylic acid were detected, including blumenol A, 3',5',5,7-tetrahydroflavones, catechins, quercetin, adenosine, salicylic acid, 2'-deoxyadenosine, p-hydroxy cinnamic acid, vanillic acid, protocatechuic acid, caffeic acid, etc. It remains to be further determined which of these components contribute greatly to bitter taste. The contribution of plant components of these components to the taste and flavor of orycho needs further study.

二月兰花粉的营养价值评价开发也是下一步的工作。

The nutritional value evaluation of orycho pollen is the next work.

二月兰的食用

二月兰叶菜干粉的生产与应用（图 2-7~ 图 2-11）

Using of orycho

Productions and application of orycho leaf dry powder（Fig. 2-7~Fig. 2-11）

图 2-7 清洗
Fig.2-7 Water cleaning

图 2-8 晾晒或干燥机烘干
Fig.2-8 Drying by airing or drying machine

图 2-9 干制二月兰叶
Fig.2-9 Orycho dry leaves

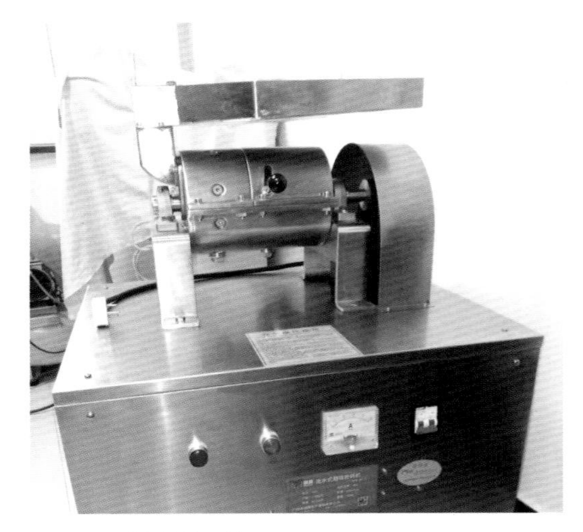

图 2-10 食品用粉碎机
Fig.2-10 Grinder for food

图 2-11 二月兰叶干粉
Fig.2-11 Orycho leaf dry powder

二月兰叶菜干粉饮品（图 2-12）。

Orycho leaf dry powder drinks (Fig.2-12).

图 2-12 二月兰叶菜干粉饮品
Fig.2-12 Orycho leaf dry powder drink

二月兰叶菜干粉食品（图 2-13~图 2-15）。

Orycho leaf dry powder foods(Fig.2-13~Fig.2-15).

图 2-13 二月兰叶菜干粉饼干

Fig.2-13 Orycho leaf dry powder biscuits

图 2-14 二月兰叶菜干粉曲奇
Fig.2-14 Orycho leaf dry powder cookies

图 2-15 二月兰叶菜干粉月饼（纪念作者上大学 40 周年）
Fig.2-15 Orycho leaf dry powder moon cake (commemorating the author's 40 th anniversary of college)

二月兰籽的食用

二月兰籽苦味饮品（图 2-16~ 图 2-17）。

Using of orycho seeds

Bitter drink of orycho seed (Fig.2-16~ Fig.2-17).

图 2-16 将二月兰籽水煮，可以做出不同浓度的二月兰籽苦味汤
Fig.2-16 The bitter soup of orycho seeds with different concentrations can be made by boiling the seeds in water

图 2-17 二月兰籽可以做复合饮品（可以加咖啡、糖、奶）
Fig.2-17 Mixed drink (coffee, sugar and milk can be added) of orycho seed

二月兰籽食用油（图 2-18）。

Edible oil of orycho seed (Fig. 2-18).

图 2-18 二月兰籽油生产工艺

Fig.2-18 Production technology of orycho seed oil

二月兰鲜叶食谱

二月兰鲜叶汉堡（图 2-19）。

Menu of fresh leaves from orycho

Hamburger made with fresh leaves from orycho (Fig.2-19).

图 2-19 二月兰叶炸鸡块汉堡
Fig.2-19 Orycho leaf / chiken meat hamburger

鲜叶炒菜（图 2-20~图 2-25）。

Stir-fry dishes (Fig.2-20~Fig.2-25).

图 2-20 二月兰炒羊肉片
Fig.2-20 Mutton slices with orycho leaf

图 2-21 二月兰炒豆腐
Fig.2-21 Tofu with orycho leaf

图 2-22 二月兰炒牛肉
Fig.2-22 Beef with orycho leaf

图 2-23 二月兰炒猪肉
Fig.2-23 Pork with orycho leaf

图 2-24 二月兰炒火腿
Fig.2-24 Ham with orycho leaf

图 2-25 二月兰摊鸡蛋
Fig.2-25 Scrambled eggs with orycho leaf

鲜叶做馅或煮汤（图 2-26~ 图 2-27）。

Leaves for the filling or soup (Fig.2-26~Fig.2-27).

图 2-26 二月兰叶饺子 / 包子 / 月饼
Fig.2-26 Jiaozi (dumplings) /Baozi/moon cake with orycho leaf

图 2-27 二月兰鸭汤
Fig.2-27 Beijing duck soup with orycho leaf

第三章 二月兰卫生学检验与毒理学评价
Hygienic inspection and toxicological evaluation of orycho

二月兰卫生学检验

对于蔬菜产品，要符合国家食品安全标准食品中污染物限量（GB 2762—2017）的要求。这些污染物是铅、镉、汞、砷、铬。我们二月兰种植基地的二月兰、水培试验的二月兰和校园内播种（图 3-1）的二月兰的检测结果都符合要求（表 3-1，图 3-2）。

Hygienic Inspection of orycho

Vegetable product should meet the requirements of national food safety standard (GB 2762-2017). These pollutants are lead, cadmium, mercury, arsenic and chromium. The detection results of orycho in the planting base, in the hydroponic experiment and in the campus (Fig.3-1) all meet the requirements (Table 3-1, Fig.3-2).

表 3-1 人工种植二月兰新鲜叶的污染物检测情况
Table 3-1 Detection of pollutants in fresh leaves of artificially planted orycho

污染物 Pollutants	食品种类 Food type	限量 Limits	种植基地 新鲜菜叶 Fresh leaf from base	种植基地 新鲜菜薹 Tender flower stalk	水培新鲜菜叶 Hydroponic fresh leaves	北大校园 新鲜菜叶 Fresh Leaves at our campus
铅 Pb（以 Pb 计）mg/kg	叶菜蔬菜	0.3	<0.05	<0.05	0.194	0.204
镉 Cd（以 Cd 计）mg/kg	叶菜蔬菜	0.2	0.0386	0.0298	0.0278	0.024
汞 Hg（以 Hg 计）mg/kg	新鲜蔬菜	0.01	<0.003	<0.003	<0.003	0.004
砷 As 总砷（以 As 计）mg/kg	新鲜蔬菜	0.5	<0.010	<0.010	<0.010	0.045
铬 Cr（以 Cr 计）mg/kg	新鲜蔬菜	0.5	<0.2	<0.2	<0.2	0.376

图 3-1 团队成员在采集自己在校园种的二月兰鲜叶
Fig.3-1 Our team members are collecting fresh leaves of orycho planted on campus

北京大学公卫学院中心仪器室测试报告

送样单位	公卫学院	送样日期	2018.10.16
样品名称	二月兰（鲜叶）	送样人	张宝旭
测试方法	ICP-MS	完成日期	2018.10.23
测试项目	Cr、Cu、Zn、As、Se、Cd、Hg、Pb		

元素	含量（ng/g）
Cr	378.37
Cu	1162.96
Zn	4504.59
As	45.49
Se	170.36
Cd	24.39
Hg	4.40
Pb	204.04

测试人员	解清	审核人	欧阳磊	负责人	崔富强

1. 样品、样品名称及送样单位均由委托单位提供，本检测报告只对委托单位所送样品负责。
2. 未经本科室书面批准，任何单位或个人不得用本报告及本中心的名义作广告宣传。

图 3-2 二月兰鲜叶的几种元素含量报告
Fig.3-2 Report on the element content of fresh leaves of orycho

二月兰的毒理学评价

虽然二月兰是我国具有 1800 多年食用历史的蔬菜，食用的区域达到十多个省份；但是，用现代毒理学技术对二月兰的安全性方面有一个比较全面的评价的话，会使我们对其价值有一个更全面的了解。

Toxicological evaluation of orycho

Orycho is a vegetable with a consumption history of more than 1800 years in China, and it is widely consumed in more than ten provinces; However, if we have a comprehensive evaluation of the safety of orycho with modern toxicology technology, we will have a more comprehensive understanding of it.

二月兰鲜叶经口急性毒性试验

试验摘要

参照《食品安全国家标准　急性经口毒性试验（GB 15193.3—2014）》对二月兰鲜叶进行了健康 ICR 小鼠的急性经口毒性试验，在试验 14 天过程中没有观察到二月兰鲜菜汁灌胃后引起动物中毒和死亡情况，二月兰鲜叶的急性半数致死量（LD_{50}）大于 160g/kg；属于实际无毒级别的食品。

Oral acute toxicity test of fresh leaves of orycho

Test summary

The acute oral toxicity test was conducted on healthy ICR mice with reference to National Standard for Food Safety—Acute Oral Toxicity Test (GB 15193.3-2014). During the 14 day test, no animal poisoning and death were observed, and the acute median lethal dose (LD_{50}) of fresh leaves of orycho was more than 160 g/kg. Orycho belongs to the actual non-toxic level.

受试物

受试物为实验人员当天在北京大学医学部校园绿地采摘的新鲜完整的二月兰叶。

将 100g 二月兰鲜叶用去离子水洗净后，加入 50mL 去离子水，用水果榨汁机打汁。

实验动物

本实验在北京大学医学部的试验设施中进行。由北京大学医学部实验动物科学部提供 SPF 级健康 ICR 小鼠 17 只（雄 8 只，雌 9 只），体重 20~23g。实验动物生产许可证号为 SCXK（京）2016-0010。实验动物设施使用许可证号为 SYXK9（京）2016-0041，动物室和实验室温度为 (23±2) ℃，自动通风，明暗周期 12h/12h，自由摄食饮水。实验动物饲料生产许可证号为 SCXK（京）2014-0010。适应环境时间大于 5 天。

方法

参照《食品安全国家标准　急性经口毒性试验（GB 15193.3—2014）》中的限量法设计了两组剂量，分别为 80g/kg 和 160g/kg[分别为 8 只动物（雌雄各半）和 9 只动物（雌 5 雄 4）]。灌胃容积分别为 20mL/kg 和 40mL/kg，两次灌胃，间隔 3h。

给受试物后几小时内严密观察动物的反应，之后每天上下午观察一次，连续观察 14 天。主要观察的指标有：动物外观、行为、分泌物、排泄物、称重、死亡情况。

Materials

The fresh and complete leaves of picked by the experimenters in the campus herb garden at Peking University Health Science Center on the same day.

100g of fresh leaves of orycho was washed with deionized water, then 50mL of deionized water was added, and the leaf juice was made with a juicer machine.

Experimental animals

This experiment was carried out in the experimental facility of Peking University Health Science Center . Seventeen SPF healthy ICR mice (8 males and 9 females), weighing 20~23g, were provided by the Department of Laboratory Animal Science, Peking University Health Science Center. Experimental animal production license No.: SCXK (Beijing) 2016-0010. The license number of experimental animal facility is SYXK9 (Beijing) 2016-0041, the temperature of animal room and laboratory is (23±2)℃ , with automatic ventilation, light and dark period of 12 h/12 h, free to eat and drink water. Production license of feed block for experimental animals is SCXK (Beijing) 2014-0010. The adaptation time is more than 5 days.

Methods

According to the limit method in National Standard for Food Safety—Acute Oral Toxicity Test (GB 15193.3−2014), two groups of doses were designed, which were 80g/kg and 160g/kg [8 animals (half male and half female) and 9 animals (5 males and 4 females) respectively]. The volume of intragastric administration was 20mL/kg and 40mL/kg, respectively, with an interval of 3 hours. The reaction of animals was closely observed within a few hours after administration, and then observed once a day in the morning and afternoon for 14 days. The main observed indexes are animal appearance, behavior, secretion, excrement, weighing and death.

试验结果

在实验期间，动物的皮肤、被毛、眼睛、呼吸、行为等未见异常，没有死亡。大体解剖未见异常。动物体重增加（图 3-3）。病理学检查未见异常。

结论

二月兰鲜叶的急性 LD_{50} 大于 160g/kg。依据《食品安全国家标准　急性经口毒性试验（GB 15193.3—2014）》规定的分级标准（LD_{50} 大于 5g/kg 为实际无毒），二月兰鲜叶属于实际无毒级别食品。

Test results

During the experiment, there was no abnormality in skin, coat, eyes, breath and behavior of animals, and no death occurred. There is no abnormality in gross anatomy. Animals gained weight (Fig.3-3). Pathological examination showed no abnormality.

Conclusion

The acute LD_{50} of fresh leaves of orycho is more than 160g/kg. According to the grading standard specified (LD_{50} greater than 5g/kg is actually non-toxic) in National Standard for Food Safety—Acute Oral Toxicity Test (GB 15193.3–2014), the fresh leaves of orycho are actually non-toxic food.

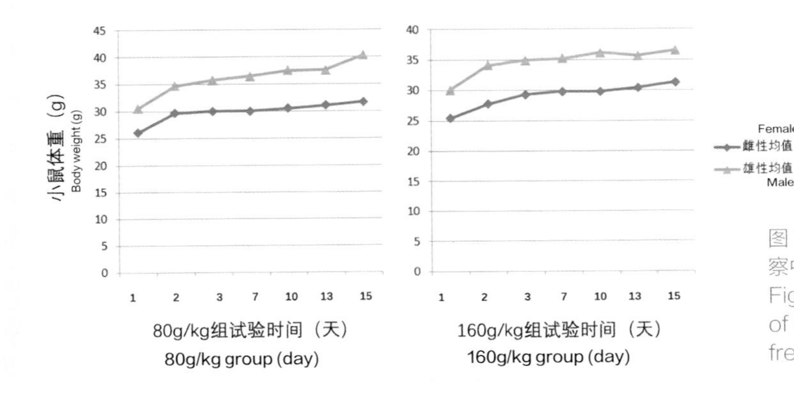

图 3-3　二月兰鲜叶急性经口试验观察中小鼠体重变化

Fig.3-3 Changes in body weight of mice during acute oral test of fresh leaves of orycho

二月兰叶干粉经口急性毒性试验

试验摘要

参照《食品安全国家标准　急性经口毒性试验（GB 15193.3—2014）》对二月兰叶干粉进行了健康ICR小鼠的急性经口毒性试验，在试验14天过程中没有观察到二月兰叶干粉灌胃后引起动物中毒和死亡情况，二月兰叶干粉的急性半数致死量（LD_{50}）大于15g/kg，属于实际无毒级别的食品。

Oral acute toxicity test of dry orycho leaf powder

Test summary

According to National Standard for Food Safety— Acute Oral Toxicity Test (GB 15193.3–2014), the acute oral toxicity test of dry leaf powder of orycho in healthy ICR mice was carried out. During the 14-day test, no animal poisoning and death caused by oral administration of dry leaf powder of orycho were observed, and the acute median lethal dose (LD_{50}) of dry powder of orycho was greater than 15g/kg, belonging to the actual non-toxic food.

受试物

受试物为二月兰叶干粉，浅绿色，房山地区采集。将细粉过 100 目筛。加入去离子水，配制成混悬液。

实验动物

本实验在北京大学医学部试验设施进行。由北京大学医学部实验动物科学部提供 SPF 级健康 ICR 小鼠 17 只（雄 8 只，雌 9 只），体重 20~23g。实验动物生产许可证号为 SCXK（京）2016-0010。实验动物设施使用许可证号为 SYXK9（京）2016-0041，动物室和实验室温度为 (23±2)℃，自动通风，明暗周期 12h/12h，自由摄食饮水。实验动物饲料生产许可证号为 SCXK（京）2014-0010。适应环境时间大于 5 天。

方法

参照《食品安全国家标准　急性经口毒性试验（GB 15193.3—2014）》中的限量法设计了 15g/kg 一次灌胃。

给受试物后几小时内严密观察动物的反应，之后每天观察一次，连续观察 14 天。主要观察的指标有：动物外观、行为、分泌物、排泄物、称重、死亡情况。

Materials

The test substance is dry powder of orycho leaf, light green, collected in Fangshan area. Sieving the fine powder with a 100 mesh sieve. Adding deionized water to prepare suspension.

Experimental animals

This experiment was carried out in Peking University Health Science Center experimental facility. Fifty SPF healthy ICR mice (8 male and 9 female), weighing 20~23g, were provided by the Department of Laboratory Animal Science, Peking University Health Science Center. Experimental animal production license No.: SCXK (Beijing) 2016-0010. The license number of experimental animal facility is SYXK9 (Beijing) 2016-0041, the temperature of animal room and laboratory is (23±2)℃ , with automatic ventilation, light and dark period of 12h/12h, free to eat and drink water. Production license of feed block for experimental animals is SCXK (Beijing) 2014-0010. The adaptation time is more than 5 days.

Methods

Dosege 15g/kg and once intragastric administration were designed according to the limit method in GB 15193.3−2014 National Standard for Food Safety Acute Oral Toxicity Test. The reaction of animals was closely observed within a few hours after administration, and then observed once a day for 14 days. The main observed indexes are animal appearance, behavior, secretion, excrement, body weight and death.

试验结果

在实验期间，动物的皮肤、被毛、眼睛、呼吸、行为等未见异常，没有死亡。大体解剖未见异常。动物体重增加见图 3-4。病理学检查未见异常。

结论

二月兰叶干粉的急性 LD_{50} 大于 15g/kg。依据《食品安全国家标准 急性经口毒性试验（GB 15193.3—2014）》规定的分级标准（LD_{50} 大于 5g/kg 为实际无毒），属于实际无毒级别食品。

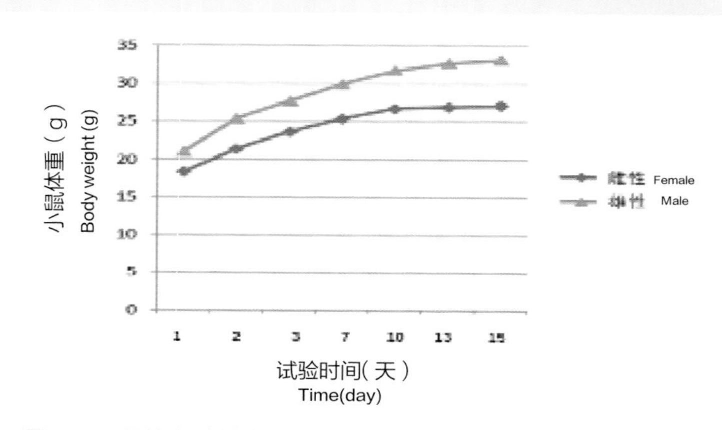

图 3-4 二月兰叶干粉急性经口试验中的小鼠体重变化
Fig.3-4 Body weight changes of mice in the acute oral test of dry powder of orycho leaves

Test results

During the experiment, there was no abnormality in skin, coat, eyes, breath and behavior of animals, and no death occurred. There is no abnormality in gross anatomy. Fig.3-4 displays the weight gain of animals. Pathological examination showed no abnormality.

Conclusion

The acute LD_{50} of the dry powder of orycho leaves is more than 15g/kg. According to the grading standard specified in National Standard for Food Safety—Acute Oral Toxicity Test (GB 15193.3-2014) (LD_{50} greater than 5g/kg is actually non-toxic), it belongs to the actual non-toxic food.

二月兰粉 28 天经口毒性试验

试验摘要

参照《食品安全国家标准　28 天经口毒性试验（GB 15193.22—2014）》对二月兰菜叶干粉（二月兰粉）进行了健康 ICR 小鼠的 28 天经口毒性试验，灌胃剂量分别为 3g/kg、6g/kg、12g/kg（最大给予量）。在试验 28 天灌胃过程中没有观察到二月兰叶干粉灌胃后死亡情况和器官病变。二月兰粉灌胃量在 6g/kg 时以上影响动物摄食及营养平衡，建议长期食用二月兰时的每日最大摄入量为干粉时不超过 3g/kg 体重（相当于鲜菜叶 30g/kg），避免造成营养失衡。

Repeated dose 28-day oral toxicity study

Test summary

According to National Standard for Food Safety—28-day oral toxicity test (GB 15193.22–2014), the oral toxicity test of dry powder of orycho leaves (orycho powder) in healthy ICR mice was carried out for 28 days, and the intragastric doses were 3g/kg, 6g/kg and 12g/kg respectively (maximum dose). During the 28-day test, no death and organ lesions were observed after the dry powder of orycho leaves was perfused. The food intake and nutritional balance of animals were affected when the oral dose of orycho powder was more than 6g/kg. It is suggested that the daily maximum intake of orycho powder should not exceed 3g/kg body weight (equivalent to 30g/kg of fresh vegetable leaves) for long-term consumption to avoid nutritional imbalance.

受试物

受试物为二月兰粉，浅绿色，房山地区采集。每 10g 鲜叶得到 1g 干粉，即 1g 干粉等于 10g 鲜叶。将细粉过 100 目筛。用去离子水配制成混悬液。

实验动物

本实验在北京大学医学部试验设施进行。由北京大学医学部实验动物科学部提供 SPF 级健康 ICR 小鼠 53 只，体重 20~22g。实验动物生产许可证号为 SCXK（京）2016-0010。实验动物设施使用许可证号为 SYXK9（京）2016-0041，动物室和实验室温度为 (23±2)℃，自动通风，明暗周期 12h/12h，自由摄食饮水。实验动物饲料生产许可证号为 SCXK（京）2014-0010。

Materials

The test substance is orycho powder, which is light green, and collected in Fangshan area. 1g of dry powder is obtained for every 10g of fresh leaves, that is, 1g of dry powder is equal to 10g of fresh leaves. Sieving the fine powder with a 100 mesh sieve. Prepare suspension with deionized water.

Experimental animals

This experiment was carried out in the experimental facility of Peking University Health Science Center. Fifty-three SPF healthy ICR mice, weighing 20~22g, were provided by the Department of Laboratory Animal Science, Peking University Health Science Center. Experimental animal production license No.: SCXK (Beijing) 2016- 0010. The license number of experimental animal facility is SYXK9 (Beijing) 2016-0041, the temperature of animal room and laboratory is (23±2)℃ , with automatic ventilation, light and dark period of 12h/12h, free to eat and drink water. Production license of feed block for experimental animals is SCXK (Beijing) 2014-0010.

方法

依据以前得到的二月兰鲜叶的急性经口 LD_{50} 大于 160g/kg，二月兰粉的 LD_{50} 大于 15g/kg，二月兰籽水煎剂的 LD_{50} 为 208.79g/kg，参照《食品安全国家标准 28 天经口毒性试验（GB 15193.22—2014）》对二月兰粉进行了健康 ICR 小鼠的 28 天经口毒性试验，灌胃剂量分别为 3g/kg、6g/kg、12g/kg（最大给予量）。每日灌胃一次，给予容量为 20mL/kg 体重，对照组给予去离子水。

观察指标：观察 28 天，试验期间至少每天观察一次动物的一般临床表现，并记录动物出现中毒的体征、程度和持续时间及死亡情况。观察内容包括被毛、皮肤、眼、黏膜、分泌物、排泄物、呼吸系统、神经系统、自主活动（流泪、异常呼吸、竖毛反应等）及行为表现（步态、姿势、对处理的反应、反常行为、有无强直性或阵挛性活动、刻板反应等）。对体质弱的动物应隔离，濒死和死亡动物及时解剖。每天灌胃时记录体重。试验结束时，计算动物体重增长量。试验结束后，进行血液学指标测定，推荐指标为白细胞计数、红细胞计数、血红蛋白浓度、血小板计数。试验结束进行血液生化指标测定，测定指标包括肝肾功能等方面。丙氨酸氨基转移酶（ALT）、乳酸脱氢酶（LDH）、血尿素氮（BUN）指标。

Methods

According to the previous data, the acute oral LD_{50} of orycho fresh leaves, orycho powder and orycho seeds decoction were more than 160g/kg, 15g/kg and 208.79g/kg, respectively. According to National Standard for Food Safety—28-day oral toxicity test (GB 15193.22–2014), the oral toxicity test of orycho powder in healthy ICR mice was carried out for 28 days, and the intragastric doses were 3g/kg, 6g/kg and 12g/kg respectively (maximum dose).The mice were given 20mL/kg body weight once a day, and the control group was given deionized water.

Observation index: Observe period was for 28 days, observation were included the general clinical manifestations of animals at least once a day during the experiment, and record the signs, degree and duration of poisoning and death of animals. The observation contents include coat, skin, eyes, mucous membrane, secretion, excreta, respiratory system, nervous system, autonomous activities (tears, abnormal breathing, vertical hair reaction, etc.) and behavior performance (gait, posture, response to treatment, abnormal behavior, whether there is tonic or clonic activity, rigid reaction, etc.). Animals with weak constitution should be isolated, and dying and dead animals should be dissected in time. Record the body weight every day during gastric perfusion. At the end of the experiment, calculate the weight gain of animals. After the test, the hematological indexes were determined. The recommended indexes are white blood cell count, red blood cell count, hemoglobin concentration and platelet count. After the test, the blood biochemical indexes were measured. The measured indexes include liver and kidney function. Alanine aminotransferase (ALT), lactate dehydrogenase (LDH) and blood urea nitrogen (BUN).

试验结果

在实验期间，动物的皮肤、被毛、眼睛、呼吸、行为等未见异常，没有死亡。动物体重增加量随剂量增加而减少（表 3-2），摄食量和食物利用率呈同样趋势，表明在剂量大于 6g/kg 以上时开始影响动物摄食和营养平衡，造成生长速度减慢。

Test results

During the experiment, no abnormality was found in the animal's skin, coat, eyes, respiration, and behavior, and there was no death. Animal body weight gain decreased with the increase in dose (Table 3-2), and food intake and food utilization showed the same trend, indicating that the dose above 6g/kg began to affect the animal's food intake and nutritional balance, resulting in decreased growth rate.

表 3-2 二月兰粉 28 天喂养后动物的体重和摄食量变化情况

Table 3-2 Effect of orycho powder on changes of body weight and food consumption in mice during repeated dose 28-day oral toxicity

小鼠 Mice	剂量 Dosage 干粉 Powder (g/kg)	剂量（相当于鲜叶）Dosage (Equivalent to fresh leaves) (g/kg)	动物数 Animals number（n）	体重增重 Body weight gain（g）	摄食量 Food consumption（g）	食物利用率 Feed conversion ratio (FCR)（%）
雌 Female	对照组 Control	对照组 Control	5	4.04±1.33	552.9	3.65
	3	30	5	3.50±1.06	459.7	3.85
	6	60	10	2.37±1.63	899.8	2.63
	12	120	6	1.63±1.62	521.3	1.84
雄 Male	对照组 Control	对照组 Control	5	15.72±2.08	711.7	11.04
	3	30	5	8.92±1.98	529.0	8.43
	6	60	10	7.55±3.03	1019.7	7.40
	12	120	7	6.77±2.66	721.5	6.57

总食物利用率 =（第 28 天体重 - 初始体重）/ 总摄食量 ×100%

Total food utilization = (Day 28 body weight - initial body weight)/total food consumption ×100%

动物器官未见明显病变，脏器系数没有明显改变（表 3-3）。

动物血液学指标测定，白细胞计数、红细胞计数、血红蛋白浓度、血小板计数没有明显改变（表 3-4）。

动物的血清生化指标测定。ALT、LDH、BUN 没有明显改变（表 3-5）。

No obvious lesion was observed in animal organs and no significant change in organ coefficients (Table 3-3).

Determination of hematological indexes in animals. There were no significant changes in hemoglobin concentration, white blood cell count, red blood cell count or platelet count in animal hematology indexes (Table 3-4).

Determination of serum biochemical indexes of animals. The ALT, LDH and BUN had no obvious changes (Table 3-5).

结论

二月兰粉的 28 天经口毒性试验显示在较短时间内重复给予没有引起动物的毒性效应，仅仅是影响动物摄食及营养平衡。当排除了动物摄食不足和营养失衡外，二月兰粉的 28 天经口毒性试验观察到有害作用剂量（NOAEL）是 12g/kg（相当于鲜叶 120g/kg）。考虑到反复食用大量二月兰可能造成身体的营养失衡问题，建议二月兰粉的使用限量为 3g/（kg · d）。

Conclusion

The 28-day oral toxicity test of orycho powder showed that repeated administration within a short period of time did not cause toxic effects on animals, but only affected their food intake and nutritional balance. When animal inadequate food intake and nutritional imbalance were excluded, the NOAEL for the 28-day period was 12g/kg (equivalent to 120g/kg fresh leaves). In view of the nutritional imbalance problem of the body possibly caused by repeatedly eating a large amount of orycho powder, it is recommended that the use limit of orycho powder is 3g/(kg·day).

表 3-3 二月兰干粉 28 天喂养后动物的脏器系数（器官重 / 体重）变化情况

Table 3-3 Effect of orycho powder on relative weight of organs in mice during repeated dose 28-day oral toxicity

小鼠 Mice	组别 (g/kg)	动物数 Animal number (n)	肝 / 体 Liver/body（%）	肾 / 体 Kindey（%）	心脏 / 体 Heart/body（%）	脾 / 体 Spleen/ body（%）	胸腺 / 体 Thyroid/ body（%）	肾上腺 / 体 Adrenal/ body（%）	肺 / 体 Lung/body（%）	性腺 / 体 Gonad/body（%）
雌 Female	对照 Control	5	4.57±0.26	1.08±0.08	0.47±0.079	0.37±0.029	0.18±0.050	0.034±0.0100	0.59±0.09	0.18±0.067
	3	5	4.83±0.17	1.17±0.07	0.42±0.050	0.38±0.037	0.19±0.068	0.029±0.0048	0.67±0.06	0.10±0.026 **
	6	10	4.71±0.24	1.15±0.05	0.48±0.045	0.37±0.081	0.18±0.079	0.027±0.0060	0.66±0.06	0.07±0.040 **
	12	6	4.77±0.51	1.20±0.08*	0.45±0.083	0.34±0.055	0.14±0.053	0.031±0.0063	0.71±0.093 *	0.09±0.026 **
雄 Male	对照 Control	5	5.30±0.38	1.41±0.10	0.46±0.041	0.26±0.035	0.11±0.021	0.021±0.0075	0.52±0.045	0.58±0.067
	3	5	5.19±0.56	1.39±0.11	0.45±0.019	0.27±0.032	0.09±0.021	0.019±0.0038	0.58±0.082	0.67±0.09
	6	10	5.01±0.47	1.41±0.17	0.47±0.026	0.30±0.047	0.10±0.038	0.018±0.0061	0.63±0.073**	0.64±0.10
	12	7	4.89±0.37	1.46±0.11	0.48±0.056	0.26±0.023	0.12±0.028	0.030±0.0081	0.62±0.047*	0.66±0.09

注：数据为均数 ± 标准差。* 与对照组相比，P<0.05；** 与对照组相比，P<0.01。性腺（雄性睾丸，雌性卵巢）。

Data presented as mean ± SEM.* Compared with the control group, P< 0.05; * * Compared with the control group, P<0.01. Gonads (male testis, female ovary).

表 3-4 二月兰干粉 28 天喂养后动物的血液学指标变化

Table 3-4 Effect of orycho powder on hematological parameters during repeated dose 28-day oral toxicity

小鼠 Mice	组别 Groups (g/kg)	动物数 Animal number (n)	血红蛋白浓度 Haemoglobin concentration (g/L)	白细胞计数 White blood cell count ($\times 10^9$/L)	红细胞计数 Red blood cell count ($\times 10^{12}$/L)	血小板计数 Platelet count ($\times 10^9$/L)
雌 Female	对照 Control	5	152±10.77	5.8±1.04	8.66±0.55	438.4±80.53
	3	5	164±13.04	7.82±1.91	9.46±0.41	392.8±122.97
	6	10	154±25.41	8.7±4.15	8.99±0.36	446.2±125.82
	12	6	155±8.03	8.4±2.33	8.63±0.36	433.3±95.15
雄 Male	对照 Control	5	152.8±5.76	4.38±0.70	8.42±0.38	447.2±76.07
	3	5	169±4.42 *	4.82±0.96	9.85±0.31 **	447.2±98.06
	6	10	164±13.51*	6.93±2.28*	9.72±0.75 **	552.3±155.03
	12	7	165.4±11.69	6.23±1.62	9.11±0.72	519.29±81.24

注：* 与对照组相比，P<0.05；** 与对照组相比，P<0.01。

* Compared with the control group, P<0.05; ** Compared with the control group, P<0.01.

表 3-5 二月兰干粉 28 天喂养后动物的小鼠血清生化指标
Table 3-5 Effect of orycho dry powder on serum biochemical parameters during repeated dose 28-day oral toxicity

小鼠 Mice	组别 Groups（g/kg）	动物数（Animal number)（n）	丙氨酸氨基转移酶 ALT（U/L）	乳酸脱氢酶 LDH（U/L）	血尿素氮 BUN（mmol/L）
雌 Female	对照 Control	5	36.2±2.95	1212.8±358.15	6.5±0.67
	3	5	34.6±0.89	961.0±240.96	6.1±0.48
	6	10	37.7±6.83	949.6±164.31	5.3±0.56 **
	12	6	42.3±14.22	909.7±224.96 *	5.4±0.60 **
雄 Male	对照 Control	5	46.6±5.13	1049.8±218.71	6.06±0.40
	3	5	38.8±7.66	1018.6±112.04	5.78±0.71
	6	10	42.0±14.77	875.6±383.59	6.10±0.88
	12	7	78.1±80.60	782.57±277.56	5.31±0.89

注：* 与对照组相比，P<0.05；** 与对照组相比，P<0.01。
* Compared with the control group, P<0.05; * *Compared with the control group, P<0.01.

二月兰粉微核试验

试验摘要

参照《食品安全国家标准　哺乳动物红细胞微核试验（GB 15193.5—2014）》，对二月兰菜叶干粉（二月兰粉）进行了健康 ICR 小鼠的经口骨髓红细胞微核的遗传毒性试验，灌胃剂量分别为 3g/kg、6g/kg、12g/kg（最大给予量）。经口给予 3g/kg、6g/kg 和 12g/kg 剂量的二月兰粉，未对小鼠骨髓细胞染色体显示损伤作用。

Micronucleus test of orycho powder

Test summary

According to National Standard for Food Safety—Mammalian Erythrocyte Micronucleus Test (GB 15193.5-2014), the genotoxicity test of oral bone marrow erythrocytes micronucleus in healthy ICR mice was conducted with dry powder of orycho leaf (orycho powder). The doses given by gavage were 3g/kg, 6g/kg and 12g/kg (the maximum dose). The oral doses of 3g/kg, 6g/kg and 12g/kg of orycho powder showed no damage to the chromosomes of mouse bone marrow cells.

受试物

受试物为二月兰粉，浅绿色，房山地区采集。每 10g 鲜叶得到 1g 干粉，即 1g 干粉等于 10g 鲜叶。将细粉过 100 目筛。用去离子水配制成混悬液。

实验动物

SPF 级 ICR 小鼠，由北京大学医学部实验动物科学部提供。动物合格证号: SCXK(京)2016-0010。动物饲养于屏障环境中，观察 3 天后用于试验。实验动物设施使用许可证号：SYXK（京）2012-0011。实验动物饲料生产许可证号为 SCXK（京）2014-0010。

Materials

The test substance is orycho powder, which is light green, and collected in Fangshan area. 1g of dry powder is obtained for every 10g of fresh leaves, that is, 1g of dry powder is equal to 10g of fresh leaves. Sieving the fine powder with a 100 mesh sieve. Prepare suspension with deionized water.

Experimental animals

SPF grade ICR mice were provided by the Department of Laboratory Animal Science, Peking University Health Science Center. Animal Certificate is SCXK (Beijing) 2016-0010. Animals were housed in a barrier environment and observed for 3 days before testing. The license of laboratory animals is SYXK (Beijing) 2012-0011. Production license of feed block for experimental animal is SCXK (Beijing) 2014-0010.

方法

选用体重 26~30g 的健康小鼠 50 只，随机分为 5 组，每组 10 只，雌、雄各半，分别为空白对照组（去离子水）、受试物高剂量组（12g/kg）、中剂量组（6g/kg）、低剂量组（3g/kg）和阳性对照组（环磷酰胺 60mg/kg）。受试物以去离子水配制成相应浓度，以经口灌胃方式给予。各受试物组、对照组及阳性对照组灌胃体积均为 0.2mL/10g。

骨髓采集和观察：第二次给药 6h 后，将小鼠脱颈椎处死，取双侧股骨，用滤纸擦掉附上股骨上的血污和肌肉，剪掉股骨头，暴露髓腔，用 1mL 注射器吸取小牛血清 0.02mL，将针头插入另一端的髓腔少许，冲洗于干净的玻片上，吹打混匀，使其成为细胞混悬液。吸取混悬液一滴于载玻片一端，按血常规法涂片，将推好晾干的标本玻片放入染色缸中用甲醇固定 15min，取出晾干。然后用新鲜配制的 10%Giemsa 染色液染色 10~15min，冲洗后晾干。

显微镜下观察涂片，每只小鼠计数 1000 个嗜多染红细胞（PCE），观察含有微核的嗜多染红细胞（PCE）数，即微核率。同时，计数嗜多染红细胞（PCE）与正染红细胞（NCE）的比例。

统计学分析：采用 SPSS 20.0 软件，按动物性别分别统计各组含微核细胞率的均数和标准差，采用泊松（poisson）分布进行统计检验，$P<0.05$ 认为差异有显著性，$P<0.01$ 认为差异有高度显著性。

试验组与对照组相比，试验结果含微核细胞率有明显的剂量－反应关系并有统计学意义时，即可确认为有阳性结果。若统计学上差异有显著性，但无剂量－反应关系时，则应进行重复试验。结果能重复可确定为阳性。

Methods

Fifty healthy mice weighing 26~30g were randomly divided into five groups, 10 for each group, with half a female and half a male, as the blank control group (deionized water), high dose group of the test object (12g/kg), medium dose group (6g/kg), low dose group (3g/kg) and positive control group (cyclophosphamide 60mg/kg). The subjects were formulated with corresponding concentrations of deionized water and administered orally. The intragastric volumes of each subject group, the control group and the positive control group were 0.2mL/10g.

Bone marrow collection and observation: Six hours after the second dose, the mice were sacrificed without cervical vertebra. The bilateral femurs were taken, and the blood stains and muscles attached to the femurs were wiped off with filter paper. The femoral head was cut off to expose the medullary cavity. 0.02mL calf serum was sucked by a 1mL syringe, and the needle was inserted into a little of the medullary cavity at the other end. After being washed on a clean glass slide, it was blown and mixed evenly to make it into a cell suspension. After one drop of suspension was sucked into one end of the slide, the slide was smeared according to the routine blood test method, and the pushed and dried sample slide was put into the staining cylinder and fixed with methanol for 15min, then taken out to dry. The sections were then stained with a freshly prepared 10% Giemsa stain for 10~15min, rinsed and allowed to dry.

The smear was observed under a microscope, and 1000 polychromatic erythrocytes (PCEs) were counted in each mouse. The number of polychromatic erythrocytes (PCEs) containing micronuclei, i.e., the micronucleus frequency, was observed. The ratio of polychromatic erythrocytes (PCEs) to positive erythrocytes (NCE) was also counted.

Statistical analysis: SPSS 20.0 software was used to calculate the mean and standard deviation of micronucleus frequency in each group according to the gender of the animals. The poisson distribution was used for statistical test. $P< 0.05$ indicated that the difference was significant, and $P<0.01$ indicated that the difference was highly significant.

Compared with the control group, a positive result was confirmed when there was a significant dose-response relationship and statistical significance of the test results including micronucleus frequency in the test group. When the difference is statistically significant but there is no dose-response relationship, repeat testing should be perfomed. Results can be repeated to determine positive.

试验结果

在实验期间，动物没有死亡。二月兰粉三个剂量组的微核细胞率与溶剂对照组比较无明显差异（$P>0.05$）；阳性对照组微核细胞率明显升高，与溶剂对照组比较差异有显著性（$P<0.01$）。雌性和雄性三个剂量组的 PCE/NCE 值与溶剂对照组比较无明显差异（$P>0.05$）（见表 3-6）。

结论

经口给予 3.0g/kg、6.0g/kg 和 12.0g/kg 剂量的二月兰粉，未对小鼠骨髓细胞染色体显示损伤作用。

Test results

No animals died during the experiment. There was no significant difference in micronucleus frequency between the three dosage groups of orycho powder and the solvent control group ($P>0.05$). The micronucleus frequency of the positive control group was significantly increased, and the difference was significant as compared with that of the solvent control group ($P<0.01$). There was no significant difference in PCE/NCE values between the female and male dose groups and the solvent control group ($P>0.05$) (Table 3-6).

Conclusion

The oral doses of 3.0g/kg, 6.0g/kg and 12.0g/kg of orycho powder showed no damage to the chromosomes of mouse bone marrow cells.

表 3-6 二月兰粉对小鼠骨髓细胞微核率的影响

Table 3-6 Effect of orycho powder on micronucleus frequency of mouse bone marrow cells

小鼠 Mice	剂量 Dose (g/kg)	动物数 Animal number (*n*)	计数细胞总数 Cell count（个）	微核率 Micronucleus frequency（‰）	PCE/NCE
雄 Male	溶剂对照 Vehicle control	5	5000	1.4±0.5	1.86±0.13
	3.0	5	5000	1.4±1.1	1.79±0.17
	6.0	5	5000	1.0±0.7	1.86±0.16
	12.0	5	5000	1.4±0.5	1.77±0.24
	阳性对照 Positive control	5	5000	19.0±1.9**	1.82±0.15
雌 Female	溶剂对照 Vehicle control	5	5000	1.2±0.4	1.80±0.28
	3.0	5	5000	0.8±0.8	1.89±0.20
	6.0	5	5000	1.2±0.8	1.86±0.28
	12.0	5	5000	1.2±0.8	1.92±0.19
	阳性对照 Positive control	5	5000	17.8±2.59**	1.76±0.29

注：** 与溶剂对照组比较，$P<0.01$。

**Compared with the solvent control group, $P<0.01$.

二月兰籽经口急性毒性试验

试验摘要

参照《保健食品检验与评价技术规范(2003版)》和《食品安全国家标准　急性经口毒性试验(GB 15193.3—2014)》对二月兰籽水煎剂进行了健康ICR小鼠的急性经口毒性试验，在试验14天过程中观察到急性半数致死量(LD_{50})为208.79g/kg，属于实际无毒级别的食品。

Oral acute toxicity test of orycho seeds

Test summary

The acute oral toxicity test of orycho seeds decoction on healthy ICR mice was conducted according to the Technical Specifications for Inspection and Evaluation of Health Foods (2003 Edition) and National Standard for Food Safety— Acute Oral Toxicity Test (GB 15193.3-2014). During the 14-day test , the acute LD_{50} was 208.79g/kg, belonging to the actual non-toxic grade of food.

受试物

二月兰籽水煎剂（产自北京），以 10 倍蒸馏水煎煮 3 次，每次 30min，滤液合并，过滤浓缩至所需量，备用。

实验动物

健康 ICR 小鼠（SPF 级），雌雄各半，体重为 20~22g，由北京大学医学部实验动物科学部提供，饲养于北京大学公共卫生学院 毒理系专门清洁动物房。实验前 5 只一笼，饲养 3d，以适应环境。动物室和实验室温度为 (23 ± 2)℃，自动通风，明暗周期 12h/12h，自由摄食饮水。饲料为北京大学医学部实验动物科学部提供的普通饲料。

Materials

The orycho seeds decoction (collected in Beijing) was decocted with 10 times of distilled water for three times, 30min each time, the filtrate was combined, filtered and concentrated to the required amount.

Experimental animals

Healthy ICR mice (SPF grade), half male and half female, weighing 20~22g, were provided by the Department of Laboratory Animal Science, Peking University Health Science Center, and reared in the special clean animal room of Department of Toxicology, Peking University School of Public Health. Before the experiment, five mice were raised in one cage for 3d to adapt to the environment. The temperature in the animal room and laboratory was (23±2)℃, with automatic ventilation, and the light and dark period was 12h/12h, with free access to food and water. The feed was ordinary feed provided by the Department of Laboratory Animal Science, Peking University Health Science Center.

方法

本试验参考《保健食品检验与评价技术规范(2003版)》进行设计。采用改良寇氏法计算LD。

预试验：在正式试验开始之前，先用少量的动物进行预试验。以找到二月兰籽水煎剂的粗略致死剂量范围，即LD_0~LD_{100}。

正式试验：健康ICR小鼠，20~22g，每组10只，雌雄各半。采用改良寇氏法进行试验。根据预试验得到的LD_0、LD_{100}以及拟分的剂量组数，按照下面的公式1，求出正式试验各剂量组组间的组距即i值，然后在LD_0~LD_{100}剂量范围内，相邻剂量组组之间按照对数等差递减设计各组的剂量，本次实验设计5个剂量组。

$$i=(\lg LD_{100}-\lg LD_0)/(n-1) \quad n\text{: 剂量组数（公式1）}$$

在给受试材之前，对小鼠禁食12h、不限制饮水。根据所设计的剂量经口给予小鼠二月兰籽水煎剂，各剂量组动物的灌胃容量相同。给材3h后恢复小鼠进食。

在给受试物后几小时内严密观察动物的反应，之后每天上下午观察一次，连续观察14d。主要观察的指标有：动物外观、行为、分泌物、排泄物、死亡情况（死亡时间、濒死前反应等）、体重变化（给材前、给材后每天均称重）。在此期间详细观察并记录动物的中毒表现和死亡情况，其中详细记录动物出现中毒情况。对所有的实验动物均进行大体解剖，包括因濒死而处死的动物、死亡的动物以及实验结束时仍存活而处死的动物。

Methods

This experiment was designed with reference to Technical Specifications for Inspection and Evaluation of Health Food (2003 edition). The LD was calculated using the modified Laplacian method.

Pre-test: Pre-test with a small number of animals before the start of the formal test. To find the approximate lethal dose range of the decoction, i.e., LD_0~LD_{100}.

Formal test: Healthy ICR mice, 20~22g, 10 in each group, half male and half female in each group. The experiment was carried out by modified Kärber method. According to the LD_0 and LD_{100} and the number of dose distance (i.e., the Formula 1) between dose groups in the formal test was calculated. Then, within the dose range of LD_0~LD_{100}, the doses of each group were designed in descending order according to logarithmic difference between adjacent dose groups. Five dose groups were designed in this experiment.

$$i=(\lg LD_{100}-\lg LD_0)/(n-1) \quad n\text{: Dosage groups (Formula 1)}$$

Mice were fasted for 12h without restriction of access to water prior to dosing. According to the designed dose, the mice were orally administered with the orycho seeds decoction, and the gavage capacity of the animals in each dose group was the same. The mice were fed again after 3h of dosing.

Animals were closely observed for reaction within a few hours of administration and once daily in the morning and afternoon for a continuous period of 14 days. The main indicators observed were: animal appearance, behavior, secretions, feces, death (time of death, pre-moribund reaction, etc.), and body weight change (the animals were weighed every day before and after feeding). During this period, the poisoning manifestations and deaths of animals were observed and recorded in detail, and the poisoning of animals was recorded in detail. Gross anatomy was performed on all experimental animals, including those that were moribund, those that died, and those that were still alive at the end of the experiment.

数据处理

根据下面的公式 2，可以求出二月兰籽水煎剂的 LD_{50}，根据公式 3 和公式 4 可以计算出二月兰籽水煎剂的 LD_{50} 的可信区间。

Data processing

According to the following Formula 2, the LD_{50} of the orycho seeds decoction can be calculated, and the confidence interval of the LD_{50} of the decoction of orycho seeds can be calculated according to Formula 3 and Formula 4.

$$\lg LD_{50}=X_m-i(\sum p-0.5)$$ Formula 2 （公式 2）

X_m：最大剂量的对数值

i：相邻两剂量比值的对数

$\sum p$：各剂量组动物死亡率总和

$$S\lg LD_{50}=i\sqrt{\sum piqi/n}$$ Formula 3 （公式 3）

pi：各组动物死亡数

$qi=1-pi$

n=各组动物数

LD_{50} 的95%可信区间 $=\lg^{-1}(\lg LD_{50}\pm S\lg LD_{50})$ Formula 4 （公式 4）

经口急性毒性试验预试验结果

预试验粗略得到二月兰籽水煎剂 LD_{100}=350g/kg（二月兰籽量），LD_0=80g/kg（二月兰籽量）。

经口急性毒性试验正式试验

根据预试验得到的二月兰籽水煎剂的粗略致死剂量范围 LD_0~LD_{100}，依据寇氏法得到剂量组间距为：

$$i=\frac{\lg LD_{100}-\lg LD_0}{4}=0.1602$$

所以二月兰籽水煎剂经口急性毒性正式试验的试验设计见表 3-7。

Results of preliminary oral acute toxicity test

The preliminary results showed that LD_{100}= 350g/kg (Raw orycho seed) and LD_0=80g/ kg(Raw orycho seed).

Oral acute toxicity test official test

According to the rough lethal dose range LD_0~LD_{100} obtained from preliminary test and the dose group spacing obtained from Karber (Korbor) method method:

$$i=\frac{\lg LD_{100}-\lg LD_0}{4}=0.1602$$

Therefore, the experimental designs for the formal oral acute toxicity test of orycho seeds decoction are shown in Table 3-7.

表 3-7 二月兰籽水煎剂经口急性毒性正式试验设计
Table 3-7 Formal experimental design for oral acute toxicity of orycho seeds decoction

小鼠 Mice	动物数 Animal number (n)	二月兰籽量 Seed dose (g/kg)	lgLD
雄性 Male	5	350.00	2.5439
雌性 Female	5		
雄性 Male	5	241.90	2.3837
雌性 Female	5		
雄性 Male	5	167.30	2.2235
雌性 Female	5		
雄性 Male	5	115.70	2.0633
雌性 Female	5		
雄性 Male	5	80.00	1.9031
雌性 Female	5		

表 3-8 二月兰籽水煎剂经口急性毒性试验小鼠死亡情况
Table 3-8 Death of mice in oral acute toxicity test of orycho seeds decoction

小鼠 Mice	二月兰籽量 Seed dose（g/kg）	死亡情况 Dead	死亡率 Dead rate (%)
雄性 Male	350.00	4/5	90
雌性 Female		5/5	
雄性 Male	241.90	4/5	50
雌性 Female		1/5	
雄性 Male	167.30	2/5	30
雌性 Female		1/5	
雄性 Male	115.70	1/5	20
雌性 Female		1/5	
雄性 Male	80.00	0/5	0
雌性 Female		0/5	

根据预试验得到的结果，正式试验分五个剂量组，分别为 350.00g/kg、241.90g/kg、167.30g/kg、115.70g/kg、80.00g/kg（生药量）。

在给予二月兰籽水煎剂后，350.00 g/kg 和 241.90g/kg 两个剂量组小鼠 2h 后出现静卧，4h 开始出现小鼠死亡（表 3-8）。

将观察过程中死亡的小鼠以及观察 14 天结束时未死亡的小鼠进行解剖，未发现小鼠脏器明显异常。

According to the results of the preliminary test, the formal test was divided into five dose groups, 350.00g/kg, 241.90g/kg, 167.30g/kg, 115.70g/kg and 80.00g/kg, respectively.

After administration of orycho seeds decoction, mice in the two dosage groups of 350.00g/kg and 241.90g/kg lay still 2h later and began to die 4h later (Table 3-8).

The dead mice during the observation and the non-dead mice at the end of observation 14d were dissected, and no obvious abnormality of mouse viscera was found.

根据死亡情况，可以进行以下计算：

$\lg LD_{50} = X_m - i(\sum_p - 0.5)$ =2.5440−0.1602(1.9−0.5)=2.31972

$LD_{50}=10^{2.31972}$ =208.79(g/kg)

$S\lg LD_{50}=i\sqrt{\frac{\sum piqi}{n}}$ =0.0427

LD_{50}的95%可信区间=$\lg^{-1}(\lg LD_{50} \pm S\lg LD_{50})$得到172.21~253.22(g/kg)

According to the mouse death, can be calculated:

$\lg LD_{50} = X_m - i(\sum_p - 0.5)$ =2.5440−0.1602(1.9−0.5)=2.31972

$LD_{50}=10^{2.31972}$ =208.79(g/kg)

$S\lg LD_{50}=i\sqrt{\frac{\sum piqi}{n}}$ =0.0427

The 95% confidence interval= $\lg^{-1}(\lg LD_{50} \pm S\lg LD_{50})$ gave values of 172.21 to 253.22(g/kg)

结论

二月兰籽的急性 LD_{50} 为 208.79g/kg。依据《保健食品检验与评价技术规范 (2003 版)》和《食品安全国家标准　急性经口毒性试验（GB 15193.3—2014）》规定的分级标准（LD_{50} 大于 5g/kg 为实际无毒），按照急性毒性分级，二月兰籽属于无毒食品。

Conclusion

The acute LD_{50} of orycho seeds was 208.79g/kg. According to the classification standards stipulated in Technical Specifications for Inspection and Evaluation of Health Food (2003 Version) and National Standard for Food Safety—Acute Oral Toxicity Test (GB 15193.3–2014) (LD_{50} > 5g/kg was considered as actual non-toxic), and according to the acute toxicity classification, orycho seed was considered as non-toxic food.

二月兰籽急性毒性和肝脏影响观察

受试物

二月兰籽水煎剂：二月兰籽 400g（产自北京），分别用 4000mL 蒸馏水煎煮 3 次，合并滤液并浓缩至 55mL，即生药量为 7.20g/mL。

实验动物

健康 ICR 小鼠（清洁级），重为 18~20g，由北京大学医学部实验动物科学部提供。实验前 5 只一笼，饲养 3d，以适应环境。动物室和实验室温度为（23±2）℃，自动通风，明暗周期 12h/12h，自由摄食饮水。饲料为北京大学医学部实验动物科学部提供的普通饲料。

方法

ICR 小鼠共 30 只，雌雄各半，随机分为对照组、两个水煎剂组，每组 10 只，雌雄各半。给药前禁食 12h，经口给予水煎剂，对照组等量的蒸馏水。第一次给药后观察 48h，随后所有动物内眦静脉取血，检测 ALT 活性。之后立即处死，解剖，取肝脏称重，观察肝脏大体情况，同时观察腹腔、胸腔其他脏器情况。使用 7170A 自动分析仪测定血清中 ALT。

用 SPSS 13.0 软件，对数据进行方差齐性检验和成组资料 t 检验处理。$P<0.05$ 认为差异有显著性，$P<0.01$ 认为差异有高度显著性。

Acute toxicity and liver effect of orycho seeds

Materials

Orycho seed 400g (from Beijing) was decocted with 4000mL distilled water for three times, respectively; the filtrates were combined and concentrated to 55mL, so that the crude drug amount was 7.20g/mL.

Experimental animals

Healthy ICR mice (clean grade), weighing 18~20g, provided by the Department of Laboratory Animal Science, Peking University Health Science Center. Before the experiment, five mice were raised in one cage for 3d to adapt to the environment. The temperature in the animal room and laboratory was (23±2)℃ , with automatic ventilation. The light and dark periods were 12h/12h, and mice were free to food and drink. The feed was ordinary feed provided by the Department of Laboratory Animal Science, Peking University Health Science Center.

Method

A total of 30 ICR mice, half male and half female, were randomly divided into a control group and two decoction groups, 10 mice in each group, half male and half female in each group. Mice were fasted for 12h before treatment and orally given decoction, while the control group was given the same amount of distilled water. The first dose was observed for 48h, followed by blood collection from the inner canthus vein in all animals for ALT activity. Immediately after that, they were sacrificed, dissected, and the liver was taken and weighed. The general condition of the liver was observed, as well as the conditions of other organs in the abdominal cavity and thoracic cavity. Serum ALT was determine using a 7170A automate analyzer.

SPSS 13.0 software was used to test the homogeneity of variance of the data and t-test the grouped data. $P<0.05$ indicated that the difference was significant, and $P<0.01$ indicated that the difference was highly significant.

试验结果

与对照组相比，二月兰籽水煎剂高剂量组雄性小鼠体重（$P<0.01$）及肝重（$P<0.05$）明显降低，但肝体比无显著变化。二月兰籽水煎剂低剂量组小鼠体重、肝重及肝体比无变化。结果见表 3-9。对存活小鼠解剖，观察各脏器变化，未发现肝脏及其他脏器有异常。

ALT 活性没有显著性改变。

Test results

Compared with the control group, the body weight and liver weight of male mice in the high dose of orycho seeds decoction group were significantly decreased ($P<0.05$), while those in the low dose of orycho seeds decoction group showed no changes. The results are shown in Table 3-9. The surviving mice were dissected and the changes of each organ were observed. No abnormality was found in the liver and other organs.

There were no significant changes in ALT activity.

表 3-9 二月兰籽水煎剂经口毒性试验结果

Table 3-9 Oral toxicity test results of orycho seeds decoction

组别 Group（g/kg）	小鼠 Mice	死亡 Death	体重 Body（g）	肝重 Liver（g）	肝 / 体 Liver/body（%）	ALT（IU/L）
对照组 Control	雄性 Male	0/5	24.35 ± 0.43	1.46 ± 0.13	6.00 ± 0.44	68 ± 16
	雌性 Female	0/5	23.3 ± 1.00	1.28 ± 0.03	5.83 ± 0.35	34 ± 17
二月兰籽 -145 Orycho seeds-145	雄性 Male	0/5	22.14 ± 3.89	1.25 ± 0.31	5.58 ± 0.43	62 ± 29
	雌性 Female	0/5	21.14 ± 1.27	1.06 ± 0.12	5.02 ± 0.38	63 ± 60
二月兰籽 -290 Orycho seeds-290	雄性 Male	2/5	17.73 ± 0.83 ##	1.12 ± 0.19 #	6.31 ± 0.78	44 ± 4
	雌性 Female	4/5	15.3	0.85	5.56	65

注：# $P<0.05$，## $P<0.01$，与对照组比较。

# $P<0.05$，## $P<0.01$，compared with the solvent control group.

结论

在本实验条件下，二月兰籽水煎剂没有观察到肝毒性。二月兰籽水煎剂的 LD_{50} 值在 290g/kg 左右。

Conclusion

Under the conditions of this experiment, no liver toxicity was observed from the orycho seeds decoction. The estimated LD_{50} of the decoction was around 290g/kg.

二月兰籽微核试验

受试物

二月兰籽水提取物（水煎剂）：二月兰籽400g（产自北京），分别用4000mL蒸馏水煎煮3次，合并滤液并浓缩至55mL，即生材量为7.20g/mL。

实验动物

本试验依据《保健食品检验与评价技术规范(2003版)》进行设计。健康ICR小鼠（清洁级），体重为18~20g，由北京大学医学部实验动物科学部提供。实验前5只一笼，饲养3d，以适应环境。动物室和实验室温度为（23±2）℃，自动通风，明暗周期12h/12h，自由摄食饮水。饲料为北京大学医学部实验动物科学部提供的普通饲料。

Micronucleus test of orycho seeds

Materials

Water extract (decoction): 400g of orycho seeds (produced in Beijing), decocted with 4000mL distilled water for 3 times respectively, combined filtrates and concentrated to 55mL, that is, the raw material amount is 7.20g/mL.

Experimental animals

This experiment was designed according to Technical Specifications for Inspection and Evaluation of Health Food (2003 Edition). Healthy ICR mice (clean grade), weighing 18~20g, were provided by the Department of Laboratory Animal Science, Peking University Health Science Center. Before the experiment, 5 animals were reared in a cage for 3 days to adapt to the environment. The temperature in animal room and laboratory is (23±2)℃ , which is ventilated automatically, with a light and dark period of 12h/12h, and free to eat and drink water. The feed is the common feed provided by the Department of Laboratory Animal Science, Peking University Health Science Center.

表 3-10 二月兰籽水煎剂对小鼠骨髓细胞微核率的影响

Table 3-10 Effect of orycho seeds decoction on micronucleus frequency of mouse bone marrow cells

分组 Groups （g/kg）	动物数 Animal number (*n*)	NCE	微核数 Micronucleus	微核率 Micronucleus frequency （‰）	PCE/NCE
对照组 Control	9	9025	6	0.665	1.802 ± 0.229
环磷酰胺 CTX	12	12108	89	7.763	1.766 ± 0.207
二月兰籽 -36 Orycho seeds-36	10	10056	5	0.497	1.767 ± 0.159
二月兰籽 -72 Orycho seeds-72	10	10050	5	0.498	1.800 ± 0.145
二月兰籽 -145 Orycho seeds-145	10	10053	4	0.398	1.797 ± 0.122

结论

微核试验是用于染色体损伤和干扰细胞有丝分裂的化学毒物的快速检测主法。一般认为微核是细胞内染色体断裂或纺锤丝受影响而在细胞有丝分裂后期滞留在细胞核外的遗传物质。微核试验能够检测化学毒物或物理因素诱导产生的染色体完整性改变和染色体分离改变这两种遗传学终点。为了评价二月兰籽的安全性，探讨其有无诱变作用，对遗传物质是否有损伤，本研究对小鼠骨髓嗜多染红细胞与正染红细胞比值、嗜多染红细胞微核率的观察，以便提供二月兰籽遗传毒理方面的资料。研究结果显示，二月兰籽各剂量组嗜多染红细胞微核率之间无显著性差异，各实验组与阴性对照组间无显著差异。二月兰籽各剂量组 PCE/NCE 无显著性差异，而二月兰籽各剂量组与阴性对照组相比，PCE/NCE 无显著性差异。本实验初步表明，二月兰籽在 145g/kg 剂量及其以下，对遗传物质没有损伤。

Conclusion

The micronucleus test is the main method for rapid detection of chemical toxicants that cause chromosome damage and interfere with cell mitosis. It is generally believed that micronucleus is the inheritance of chromosome breakage in cells or the influence of spindle on the retention of micronucleus outside the nucleus in the late mitosis of cells. The micronucleus assay is capable of detecting two genetic endpoints, changes in chromosome integrity and changes in chromosome segregation, induced by chemical toxicants or physical factors. In order to evaluate the safety of orycho seed and to explore whether it had mutagenic effect or not and whether it had damage to genetic materials, the ratio of polychromatic erythrocytes to normal erythrocytes in mouse bone marrow and the micronucleus frequency of polychromatic erythrocytes were observed in this study in order to provide information on genetic toxicity of orycho seed. The results showed that there was no significant difference in micronucleus frequency of polychromatic erythrocytes between different dosage groups of orycho and between different experimental groups and negative control group. There was no significant difference in PCE/NCE among different dosage groups of orycho seeds, while compared with the negative control group, there was no significant difference in PCE/NCE among different dosage groups of orycho seeds. The preliminary results of this experiment showed that there was no damage to genetic material at the dose of 145g/kg or below.

二月兰籽致畸试验

摘要

（1）目的

在小鼠胚胎器官形成期，通过经口灌胃给予孕小鼠二月兰籽水煎剂，观察二月兰籽对胎小鼠的影响，评价二月兰籽对小鼠有无致畸作用。

（2）方法

在小鼠受孕 7~12 天维生素 AD 组经口灌胃给予 6.25g/kg 的维生素 AD；在小鼠受孕 6~15 天阴性对照组经口灌胃给予蒸馏水，各剂量组经口灌胃给予低、中、高三个剂量组（13.0g/kg，26.0g/kg，52.0g/kg）的二月兰籽水煎剂，在受孕 18 天处死孕小鼠，观察二月兰籽对孕小鼠及胎小鼠的影响。

（3）结果

各剂量组胎小鼠的形成和发育、雌雄比例、外观畸形、骨骼和内脏畸形与阴性对照组比较，其差异均无统计学意义（$P>0.05$）。

（4）结论

二月兰籽对小鼠无致畸性。

Teratogenicity study of orycho seeds

Abstract

(1) Objective

To study the teratogenic effect of orycho seed decoction in mice. The mice were administrated orally with orycho seeds decoction during the formation time of embryonic organ.

(2) Methods

6.25g/kg vitamin AD was given orally in the vitamin AD group 7~12 days after conception, distilled water was given orally in the negative control group 6~15 days after coneption, and the low, medium and high dose groups (13.0g/kg, 26.0g/kg and 52.0g/kg) were given orally in each dose group, and the pregnant mice were killed at 18 days after conception.

(3) Results

In the study, compared with the negative control group, no statistically significant difference was found in each dosage group ($P>0.05$), including the formation and development of embryo, proportion of female and male, surface blemish, skeletal and internal organ anomaly.

(4) Conclusion

In the study, there was no teratogenic effect of orycho seed in mice.

受试物

二月兰籽（产自北京），以 10 倍蒸馏水煎煮 3 次，每次 30min，滤液合并，过滤浓缩至所需量，备用。

维生素 AD 滴剂：广州珠江制药厂，规格为每克含维生素 A 50000 单位，批号为 110301。

实验动物

性成熟 8 周龄健康 ICR 小鼠（SPF 级），雄性体重 30~32g，雌性体重 25~27g，由北京大学医学部实验动物科学部提供，饲养于北京大学公共卫生学院毒理系专门清洁动物房。实验前 5 只一笼，饲养 3 天，以适应环境。动物室和实验室温度为 (23±2)℃，自动通风，明暗周期 12h/12h，自由摄食饮水。饲料为北京大学医学部实验动物科学部提供的普通饲料。

Materials

The seeds of orycho (produced in Beijing) were decocted with 10 times of distilled water for three times, 30min each time, the filtrate was combined, filtered and concentrated to the required amount for standby.

Vitamin AD drops: Guangzhou Zhujiang Pharmaceutical Co., Ltd., with the specification of 50000 units per gram of vitamin A, Batch No.110301.

Experimental animals

Healthy ICR mice (SPF grade) with 8 weeks of sexual maturation, weighing 30~32g in males and 25~27g in females, were provided by the Department of Laboratory Animal Science, Peking University Health Science Center, and reared in the special clean animal room of Department of Toxicology, Peking University School of Public Health. Before the experiment, five mice were raised in one cage for 3 d to adapt to the environment. The temperature in the animal room and laboratory was (23±2)℃, with automatic ventilation, and the light and dark period was 12h/12h, with free access to food and water. The feed was ordinary feed provided by the Department of Laboratory Animal Science, Peking University Health Science Center.

方法

本试验依据《保健食品检验与评价技术规范(2003版)》进行设计，设阴性组、阳性组（维生素AD组）、3个剂量组，每个剂量组的剂量依据经口急性毒性试验得到的LD_{50}来确定，为1/32LD_{50}、1/16LD_{50}、1/4 LD_{50}，即13g/kg、26g/kg、52g/kg（二月兰籽）。另设阴性对照组和阳性对照组，其中阴性对照组动物给予蒸馏水，阳性对照组动物给予6.25g/kg维生素AD滴剂（相当于312500IU/kg）。

本试验采用交配后分组方法，将已交配的孕小鼠随机分配到各个试验组，保证每组至少有孕小鼠20只。

（1）受孕动物的检查和受试物时间

将性成熟雌雄小鼠按照1：1比例于每天晚上7:00进行合笼，次日早晨7:00观察雌小鼠阴栓，查出阴栓，则认为该雌小鼠已经交配，当日记作“受孕”0天。检出的“孕小鼠”随机分到各组，并称重和编号，在受孕6~15天经口给予受试物（阴性对照组给予蒸馏水），维生素AD组小鼠在第8~12天经口给予维生素AD。在受孕的第0、3、6、9、12、15、18天称重，并计算相应的给药量。

（2）孕小鼠处死和检查

于小鼠受孕第18天直接断头处死，剖腹、完整的取出子宫以及卵巢并称重，记录并检查黄体数、吸收胎数、早死胎数、晚死胎数以及活胎数。

Methods

This experiment was designed according to Technical Specifications for Inspection and Evaluation of Health Food (2003 Version). A negative group, a positive group (vitamin AD group) and three dose groups were set up. The dose of each dose group was determined according to the LD_{50} obtained from the oral acute toxicity test, which was 1/32, 1/16 and 1/4 LD_{50}, namely 13g/kg, 26g/kg and 52g/kg (orycho seed). In addition, there were a negative control group and a positive control group. The animals in the negative control group were given distilled water and the animals in the positive control group were given 6.25g/kg vitamin AD drops (equivalent to 312,500IU/kg).

In this study, the mated pregnant mice were randomly assigned to each experimental group using the post- mating grouping method to ensure that there were at least 20 pregnant mice in each group.

(1) Examination and test article time of pregnant animal

The sexually mature male and female mice were cooped up at 7:00 p.m. every day in a 1：1 ratio. The female mouse negative thrombi were observed at 7:00 a.m. the next day. If they were found, the female mice were considered to have mated. The diary was used for "conception" for 0 d. The detected "pregnant mice" were randomly assigned to each group, weighed and numbered. The mice were orally administered with the test substance (distilled water for the negative control group) within 6–15 days of conception, and the mice in the vitamin AD group were orally administered with vitamin AD within 8–12 days. The animals were weighed on days 0, 3, 6, 9, 12, 15, and 18 of conception and the appropriate dose was calculated.

(2) Sacrifice and examination of pregnant mice

On the 18th day of pregnancy, the mice were sacrificed by direct decapitation, and laparotomy. The uterus and ovaries were completely removed and weighed. The number of corpora lutea, number of absorbed fetuses, number of early and late stillbirths, and number of live fetuses were recorded and examined.

（3）活胎数检查

如图 3-5 所示，将所有胎小鼠进行编号，称取胎盘以及胎小鼠重量。详细观察胎小鼠性别，用游标卡尺测量小鼠体长、尾长，仔细检查小鼠外观有无异常。

对每一只胎小鼠的头部、躯干、四肢的外观都按照保健食品检验与评价技术规范对于致畸试验的活胎数检查的技术规范进行详细的观察。致畸试验胎小鼠体表检查项目见表 3-11。

(3) Live tire count check

As shown in Fig.3-5, all fetal mice were numbered, and the placenta and fetal rat weight were taken. The sex of the fetus was observed in detail, and the body length and tail length were measured with a vernier caliper. The appearance of the mice was examined carefully for abnormalities.

The appearance of head, trunk and limbs of each fetal rat was observed in detail in accordance with the Technical Specifications for Inspection and Evaluation of Health Food and the Technical Specifications for Live Fetal Number Examination in the Teratogenic Test. The body surface examination items of fetal mice tested in the sensitive teratogenic period are shown in Table 3-11.

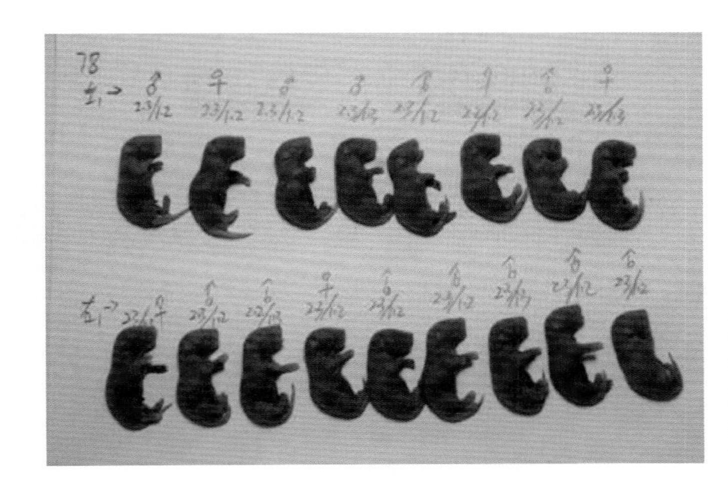

图 3-5 活胎外观检查（将胎小鼠按照其在子宫中的位置进行排序，详细检查外观）
Fig.3-5 live tire appearance inspection (Fetal mice were sequenced by their position in the uterus and examined for appearance in detail)

表 3-11 胎小鼠体表检查项目
Table 3-11 Body surface examination of fetal mice

头部 Head	躯干部 Trunk	四肢 Arms and legs
无脑症 Anencephaly	胸骨裂 Sternal cleft	多肢 Multiple limbs
脑膨出 Encephalocele	胸部裂 Thoracic fissure	无肢 Limb-free
头盖裂 Split head cover	脊柱裂 Spina bifida	短肢 Short limb
脑积水 Hydrocephalus	脊柱侧弯 Scoliosis	半肢 Half limb
小头症 Microcephaly	脊柱后弯 Kyphoscoliosis	多指 Polydactylism
颜面裂 Facial cleft	脐疝 Umbilical hernia	无指 No finger
小眼症 Microphthalmia	尿道下裂 Hypospadias	并指 Syndactylia
眼球突出 Protopsis	无肛门 No anus	短指 Brachydactylia
无耳症 Auricular disease	短尾卷尾 Short tail	缺指 Missing finger
小耳症 Small ear	无尾 Tailless	
耳低位 Ear low	腹裂 Gastroschisis	
无颚症 Jawless		
小颚症 Micrognathia		
下颚裂 Mandibular fissure		
口唇裂 Oral cleft lip		

（4）胎小鼠骨骼标本的制作与检查

a. 编号：将每窝胎小鼠按照其在子宫内的着床顺序（从右子宫角至左子宫角）进行编号。

b. 固定：选取奇数编号的胎小鼠放入装有95% 乙醇的平皿中固定 10min。

c. 去皮、去脂肪、去内脏：将胎小鼠从95% 乙醇中取出，用镊子除去胎小鼠皮肤、内脏以及背部肩胛骨的脂肪。

d. 腐蚀：放入 1% 氢氧化钾溶液中腐蚀24h。

e. 染色：将胎小鼠从氢氧化钾溶液中取出，放入茜素红溶液中染色 24~48h，染色当天摇动玻璃瓶 2~3 次，待骨骼染成红色为止。

f. 透明：将胎小鼠换入透明液 A 中 24h，然后换入透明液 B 中 24~48h，待胎小鼠骨骼部分已染红，而软组织的紫红色基本褪去。

g. 保存：将胎小鼠置换入甘油中保存。

制作完成的胎小鼠骨骼标本如图3-6所示，可以清楚地看到胎小鼠的骨骼发育情况。

对于已经制作完成的胎小鼠骨骼标本，要按照从上到下的顺序详细观察骨骼的发育情况。胎小鼠骨骼检查项目见表 3-12。

(4) Preparation and examination of fetal mouse bone specimen

a. Numbering: Each litter of fetal mice was numbered according to the sequence of their implantation in the uterus (from the right uterine horn to the left uterine horn).

b. Fixation: The odd-numbered fetal mice were fixed in a plate containing 95% ethanol for 10min.

c. Peeling, removing fat and viscera: the fetal mice were taken out from 95% ethanol, and the fat in the skin, viscera and back scapula of the fetal mice was removed with forceps.

d. Corrosion: the mixture was put into 1% potassium hydroxide solution for corrosion for 24h.

e. Staining: the fetal mice were taken out from the potassium hydroxide solution and put into the alizarin red solution for staining for 24–48h. On the day of staining, the glass bottle was shaken for 2–3 times until the bones were dyed red.

f. Transparency: The fetal mice were changed into the transparent liquid A for 24h, and then into the transparent liquid B for 24–48h, until the skeletal parts of the fetal mice were dyed red, and the purplish red of the soft tissues almost faded.

g. Preservation: the fetal mice were replaced by glycerol for preservation.

The prepared fetal rat bone specimen is shown in Fig. 3-6 and the bone development of the fetal mouse can be clearly seen.

For the prepared fetal mouse bone specimens, the bone development shall be observed in detail from top to bottom. The fetal mouse bone examination items are shown in Table 3-12.

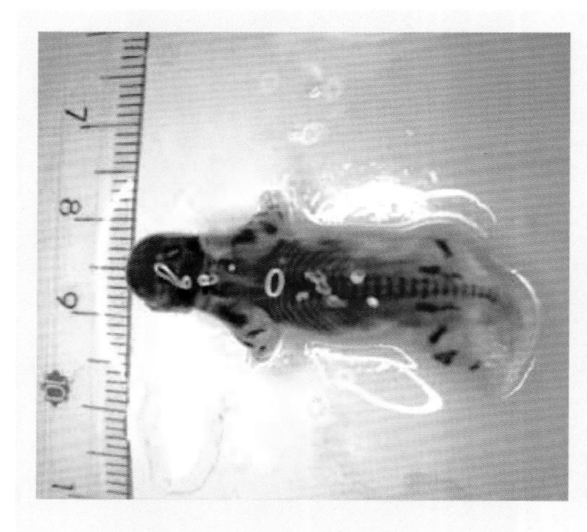

图 3-6 胎小鼠骨骼标本（制作完成的骨骼标本可以清楚地观察到胎小鼠骨骼的发育情况）
Fig.3-6 Fetal mouse bone specimen (The development of fetal mouse bones can be clearly observed with the prepared bone specimen)

表 3-12 致畸试验胎小鼠骨骼检查项目
Table 3-12 Skeletal examination items of teratogenic test fetus mice

骨骼 Skeleton	项目 Items
枕骨 Occipital bone	骨化中心、发育不全 Ossification center, hypoplasia
脊柱骨 Spinal bone	数目、融合、纵裂、部分裂开、骨化中心数、发育不全、缩窄、脱离、形状 Number, fusion, longitudinal fissure, partial fissure, number of ossification centers, hypoplasia, narrowing, detachment, shape
骨盆 Basin	弓数目、骨化中心数、性状异常、融合、裂开、缩窄、脱离 Number of arches, number of ossification centers, abnormal character, fusion, dehiscence, narrowing, detachment
四肢骨 Limb bone	形状、数目 Shape, number
腕骨 Carpus	骨化中心数 Number of ossification centers
掌骨 Metacarpal bone	形状 Shape
趾骨 Phalanx	形状 Shape
肋骨 Rib	数目、形状、融合、分叉、缺损、发育不全 Number, shape, fusion, bifurcation, defect, hypoplasia
胸骨 Sternum	形状、完全缺损、胸骨节融合、裂开、形状异常、发育不全 Shape, complete defect, sternal fusion, dehiscence, abnormal shape, hypoplasia

（5）胎小鼠内脏检查

每窝胎小鼠按照其在子宫内的着床顺序（从右子宫角至左子宫角）进行编号，其中选取偶数编号的胎小鼠放入 Bouins 液中固定 2 周，做内脏检查。

（6）数据处理

孕小鼠的增重、子宫连胎重、胎小鼠身长、胎小鼠尾长、胎盘重、胎小鼠重等计量资料采用单因素方差分析；活胎仔数、黄体数、着床数、等计数资料，使用卡方检验。胎数的数据以窝为单位进行统计。

(5) Visceral examination of fetal mice

Each litter of fetal mice was numbered according to the order of their implantation in the uterus (from the right uterine horn to the left uterine horn). Even-numbered fetal mice were fixed in Bouins solution for 2 weeks and underwent visceral examination.

(6) Data processing

The one-way analysis of variance was used to analyze the measurement data including pregnant mouse weight gain, uterine fetal weight, fetal mouse length, fetal mouse tail length, placental weight, and fetal mouse weight. The chi-square test was used to analyze the count data of live fetuses, corpus luteum, implantation, and others. The number of fetuses was counted in litters.

试验结果

（1）孕小鼠体重变化结果

各组孕小鼠体重变化统计结果见表 3-13，对各组孕小鼠在受孕第 0 天、第 3 天、第 6 天、第 9 天、第 12 天、第 15 天、第 18 天的体重以及孕小鼠增重情况进行比较，未发现明显异常，组间比较差异无统计学意义（*P*>0.05）。

（2）孕小鼠检查结果

各组孕小鼠检查结果见表 3-14，对各组孕小鼠的黄体数、活胎数、吸收胎数、早死胎数、晚死胎数进行分析比较，显示维生素 AD 组活胎数低于阴性组、晚死胎数高于阴性组，其差异具有统计学意义（*P*<0.05），二月兰籽各剂量组与阴性组比较，差异均无统计学意义（*P*>0.05）。维生素 AD 组出现 3 只吸收胎，其他组未观察到吸收胎。

（3）胎小鼠雌雄以及外观畸形观察结果

胎小鼠外观检查观察到维生素 AD 组胎小鼠出现明显的短尾、短肢、卷尾、张耳、开眼等畸形，其余各组未观察到类似情况（表 3-15）。

Test results

(1) Results of body weight changes pregnant mice

The statistical results of body weight changes of pregnant mice in each group are shown in Table 3-13. The body weights of pregnant mice on the 0th, 3rd, 6th, 9th, 12th, 15th and 18th days of conception as well as the body weight gains of pregnant mice in each group were compared. No obvious abnormality was found, and the difference between groups was not statistically significant (*P*>0.05).

(2) Examination results of pregnant mice

The examination and statistical results of pregnant mice in each group are shown in Table 3-14. The analysis and comparison of the corpus luteum number, live fetuses number, absorbed fetuses number, premature fetal death number and late fetal death number of pregnant mice in each group showed that the live fetus number in the vitamin AD group was lower than the negative group, and the late dead fetus number was higher than the negative group. The differences were statistically significant (*P*<0.05). There was no significant difference between each dosage group of orycho seed and the negative group (*P*>0.05). Three absorption fetuses were observed in the vitamin AD group, while no absorption fetuses were observed in the negative control group and in either dose group of orycho.

(3) Male and famale fetuses of each group and appearance deformity

In the external examination of fetal mice, it was observed that the fetal mice in the vitamin AD group had obvious malformations such as short tail, short limb, coil tail, open ears, and open eyes, etc., while no similar conditions were observed in the other groups (Table 3-15).

表 3-13 各组孕小鼠体重变化结果
Table 3-13 Body Weight of Pregnant Mice

分组 Group (g/kg)	孕鼠数 Pregnant mice （*n*）	第 0 天 Day 0 （g）	第 3 天 Day 3 （g）	第 6 天 Day 6 （g）	第 9 天 Day 9 （g）	第 12 天 Day 12 （g）	第 15 天 Day 15 （g）	第 18 天 Day 18 （g）	增重 Body Gain （g）
对照组 Control	17	31.8±3.2	33.6±3.1	35.4±3.1	38.1±3.3	44.7±4.4	55.7±7.1	68.8±7.3	37.0±4.6
维生素 AD Vitamin AD	18	32.4±3.1	34.4±3.4	36.4±3.4	39.2±3.5	43.3±4.5	54.2±6.1	69.0±8.7	36.6±6.6
二月兰籽 -13 Orycho seeds-13	19	31.3±2.0	33.7±2.6	35.5±2.7	37.8±2.7	44.5±3.6	54.1±4.3	68.2±7.1	36.8±5.8
二月兰籽 -26 Orycho seeds-26	17	30.0±1.2	32.7±1.8	34.7±1.7	36.1±1.7	43.2±2.2	53.3±2.3	66.3±3.7	36.3±3.0
二月兰籽 -52 Orycho seeds-52	19	30.6±2.0	33.2±2.3	35.0±2.3	37.3±2.5	43.6±2.9	53.8±4.2	66.2±6.5	35.6±5.6

表 3-14 各组孕小鼠检查结果
Table 3-14 Examination results of pregnant mice

分组 Group (g/kg)	孕鼠数 Pregnant mice (n)	黄体数 Corpus luteum (n)	活胎数 Live fetuses (n)	吸收胎数 Absorbed fetuses (n)	早死胎数 Premature fetal death (n)	晚死胎数 Late fetal death (n)
对照组 Control	17	309	266	0	8	7
维生素 AD Vitamin AD	18	315	234	3	$15^{\#}$	$35^{\#}$
二月兰籽 -13 Orycho seeds-13	19	329	287	0	9	4
二月兰籽 -26 Orycho seeds-26	17	295	253	0	5	5
二月兰籽 -52 Orycho seeds-52	19	313	270	0	7	5

注：$^{\#}P < 0.05$（维生素 AD 组与对照组比较）。
$^{\#}$Compared with control group, $P<0.05$.

表 3-15 各组胎小鼠雌雄以及外观畸形观察结果

Table 3-15 Male and female fetuses of each group and appearance deformity

分组 Group (g/kg)	活胎数 Live fetuses （n）	雌性胎小鼠数 Fetus female （n）	雄性胎小鼠数 Fetus male （n）	雌雄比例 Female/Male	短尾胎小鼠数 Short tail （n）	短肢胎小鼠数 Short limb （n）	张耳胎小鼠数 Open ears （n）	开眼胎小鼠数 Open eye （n）	卷尾胎小鼠数 Coil tail （n）
对照组 Control	266	113	153	0.739	0	0	0	0	0
维生素 AD Vitamin AD	234	122	113	1.080#	152	58	165	21	50
二月兰籽 -13 Orycho seeds-13	283	122	161	0.758	0	0	0	0	0
二月兰籽 -26 Orycho seeds-26	253	114	147	0.775	0	0	0	0	0
二月兰籽 -52 Orycho seeds-52	270	119	151	0.788	0	0	0	0	0

注：$^{\#}P < 0.05$（维生素 AD 组与阴性组比较）。

$^{\#}P<0.05$ (Compared with control group).

（4）胎小鼠外观检查比较

各组胎小鼠外观检查比较如图3-7所示，维生素AD组观察到部分胎小鼠有开眼现象，其胎小鼠四肢、尾长均小于阴性组。而二月兰籽各剂量组与阴性组比较，未观察到明显变化（表3-16）。

(4) Comparison of external examination of fetal mice

As shown in Fig.3-7, the appearance examination comparison of fetal mice in each group showed that some fetal mice in the vitamin AD group had eyes opening phenomenon, and the lengths of four limbs and tail of the fetal mice were smaller than those in the negative group. No significant changes were observed in different dosage groups of orycho compared with the negative group (Table 3-16).

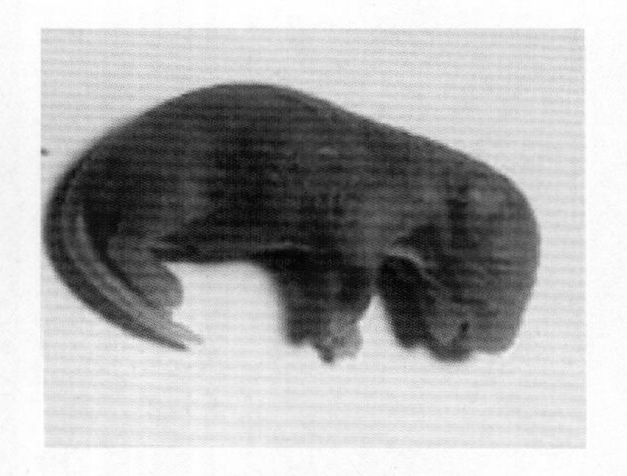

对照组
Control

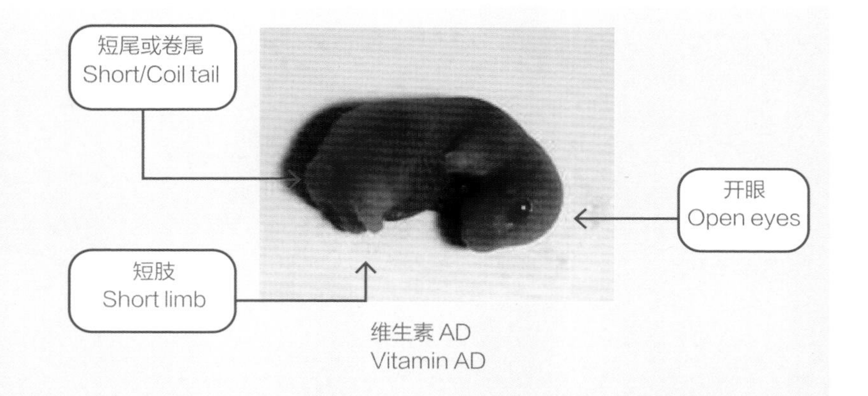

维生素 AD
Vitamin AD

二月兰籽 -13
Orycho seeds-13

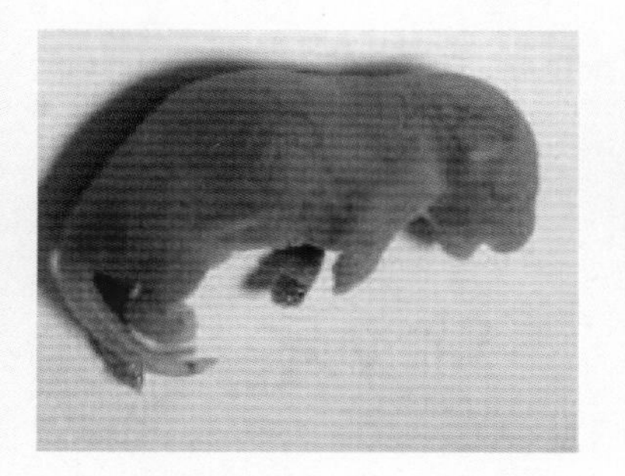

二月兰籽 -26
Orycho seeds-26

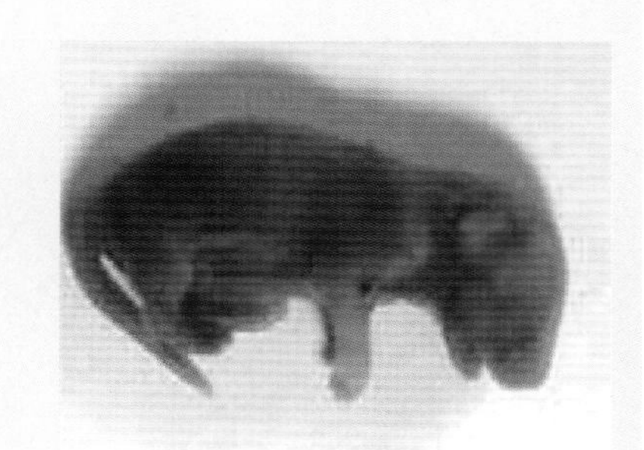

二月兰籽 -52
Orycho seeds-52

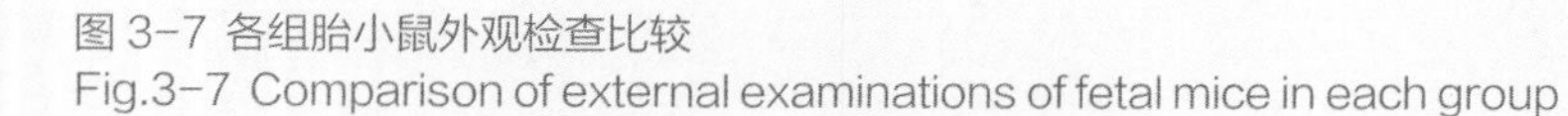

图 3-7 各组胎小鼠外观检查比较
Fig.3-7 Comparison of external examinations of fetal mice in each group

表 3-16 各组胎盘重、胎小鼠体重、身长及尾长情况
Table 3-16 Placental weight, fetal mouse weight, length and tail length in each group

分组 Group (g/kg)	孕小鼠数 Pregnant mice (n)	胎盘重 Placental weight (g)	胎小鼠体重 Fetal mouse weight (g)	胎小鼠身长 Fetal mouse length (cm)	胎小鼠尾长 Fetal mouse tail length (cm)
对照组 Control	17	0.104±0.019	1.35±0.11	2.36±1.16	1.225±0.120
维生素 AD Vitamin AD	18	0.123±0.023#	1.36±0.25	2.27±0.12	0.760±0.294#
二月兰籽 -13 Orycho seeds-13	19	0.108±0.050	1.33±0.11	2.34±1.24	1.227±0.053
二月兰籽 -26 Orycho seeds-26	17	0.108±0.018	1.34±0.10	2.26±0.08	1.221±0.085
二月兰籽 -52 Orycho seeds-52	19	0.103±0.016	1.32±0.15	2.26±0.09	1.219±0.059

注：# $P<0.05$（与对照组比较）。
# $P < 0.05$ (Compared with control group).

（5）胎小鼠骨骼检查

选取胎骨骼检查中有代表性的畸形情况进行分析，胎小鼠骨骼检查结果见表 3-17。由结果可以看出，维生素 AD 组出现畸形的胎小鼠数量明显多于对照组，其主要畸形为囟门异常、肋骨出现 14 肋、四肢骨、趾骨以及尾椎异常。维生素 AD 组出现畸形的胎小鼠数量占本组观察胎小鼠的 71.19%，与对照组比较，差异具有统计学意义（$P<0.05$）。各剂量组出现畸形胎小鼠数量占本组观察胎小鼠数量的百分比与对照组比较差异无统计学意义（$P>0.05$）。

胎小鼠骨骼观察可以明显地看到维生素 AD 组胎小鼠的囟门偏大、四肢骨短小、趾骨缺失、尾椎发育不良以及肋骨出现 14 肋等现象，其余各组只是偶尔见到这些畸形。详情见图 3-8~ 图 3-11。

(5) Skeletal examination of fetal mice

Representative malformations in fetal bone examination were selected and analyzed. The results of fetal mouse bone examination are shown in Table 3-17. From the results, we could see that the number of fetal mice with malformations in the vitamin AD group was significantly greater than that in the negative group. The main malformations were fontanel abnormalities, 14th rib occurred, limbs bone, digital bone and caudal vertebra abnormalities. The number of malformed fetal mice in the vitamin AD group accounted for 71.19% of the observed fetal mice in this group, and the difference was statistically significant as compared with the negative group ($P<0.05$). The percentage of the number of malformed fetuses in each dose group in the number of observed fetuses was not significantly different from that in the blank control group ($P>0.05$).

The skeletal observation of fetal mice in the vitamin AD group revealed obvious phenomena such as large fontanelle, short bones of four limbs, missing digital bone, dysplasia of caudal vertebra and appearance of 14th rib in ribs. These malformations were only occasionally seen in other groups. See Fig.3-8~Fig.3-11 for details.

表 3-17 胎小鼠骨骼畸形胎发生率（%）
Table 3-17 Fetal rate of skeletal deformity in fetal mice (%)

分组 Group (g/kg)	囟门 Fontanel（%）	枕骨 Occipital bone（%）	颈椎 Cervical vertebra（%）	肋骨（14 肋）Ribs (14 ribs)（%）	肋骨（变形）Rib (deformation)（%）	胸椎 Sternal vertebra（%）	腰椎 Lumbar vertebra（%）	胸骨 Sternum（%）	四肢骨 Limb bone（%）	趾骨 Digital bone（%）	骨盆 Basin（%）	尾椎 Caudal vertebra（%）	合计 Total（%）
对照组 Control	0.8	0.00	0.00	2.27	0.80	0.00	0.00	0.00	1.51	3.78	0.00	0.80	9.85
维生素 AD Vitamin AD	67.95$^{\#}$	0.00	1.69	27.12$^{\#}$	0.00	2.54	11.01	0.00	66.10$^{\#}$	65.25$^{\#}$	0.00	67.80$^{\#}$	71.19$^{\#}$
二月兰籽 -13 Orycho seeds-13	0.00	0.00	0.00	1.43	0.00	0.00	0.00	0.00	1.43	5.71	0.00	0.71	9.29
二月兰籽 -26 Orycho seeds-26	0.00	0.00	0.00	4.76	1.59	0.79	0.00	0.00	0.00	2.38	0.00	0.00	9.09
二月兰籽 -52 Orycho seeds-52	0.00	0.00	0.00	7.97	0.72	0.00	0.00	0.00	0.00	1.45	0.00	0.00	10.14

注：$^{\#}P < 0.05$（与对照组比较）。
$^{\#}P < 0.05$ (Compared with control group).

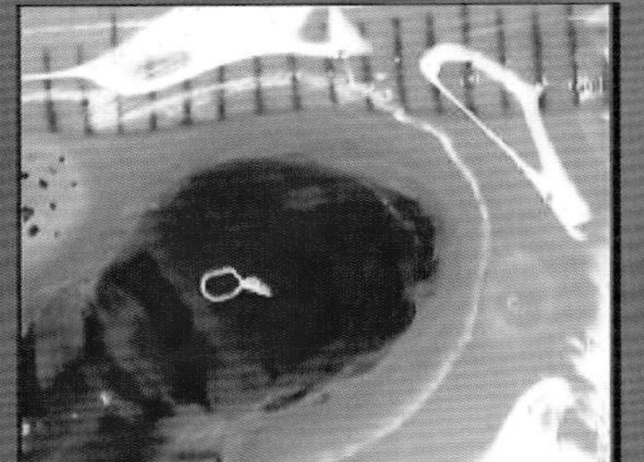

对照组
Control

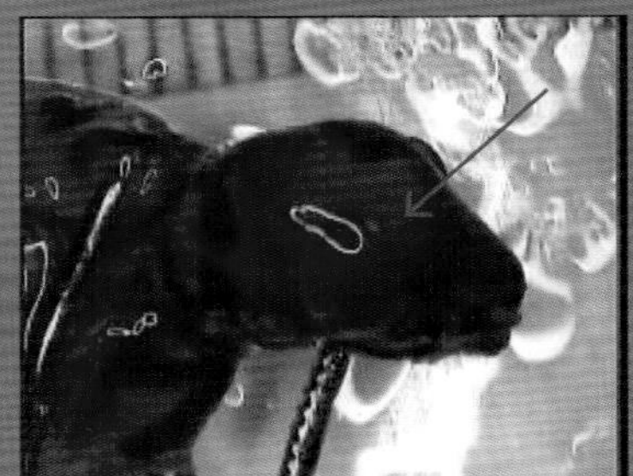

维生素 AD
Vitamin AD

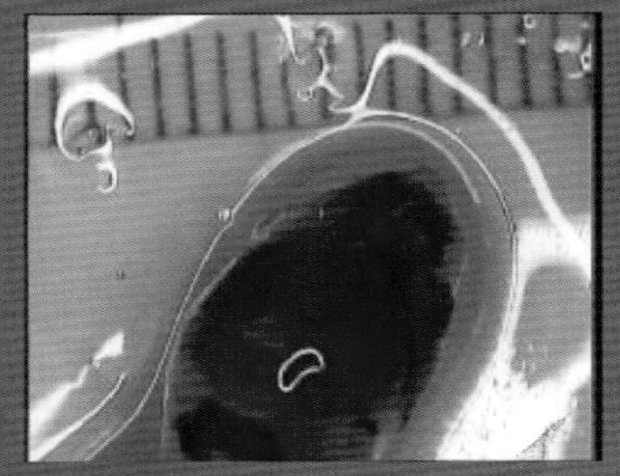

二月兰籽 -13
Orycho seeds-13

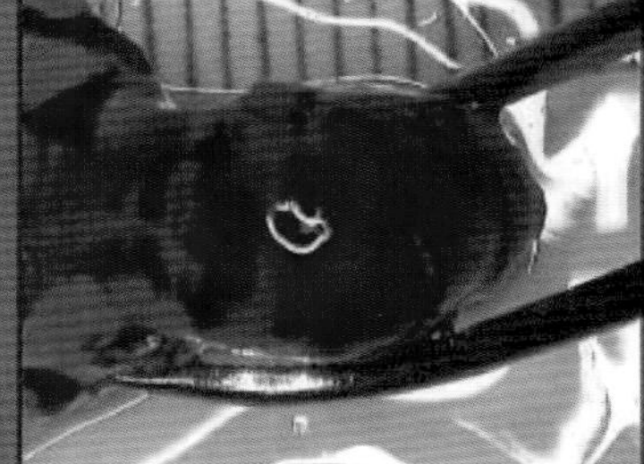

二月兰籽 -26
Orycho seeds-26

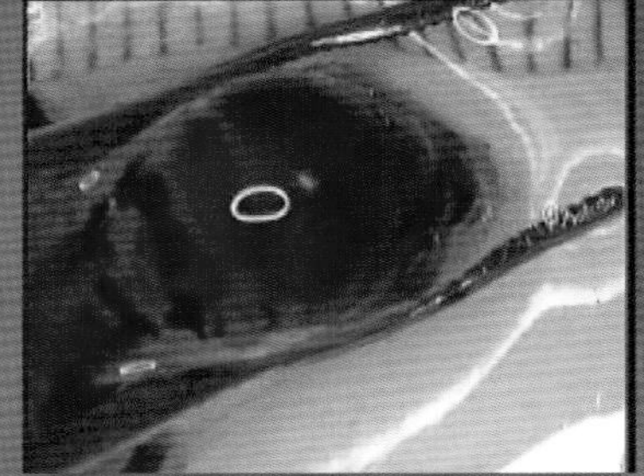

二月兰籽 -52
Orycho seeds-52

图 3-8 胎小鼠囟门检查比较
Fig.3-8 Comparison of fontanel examination of fetal mice

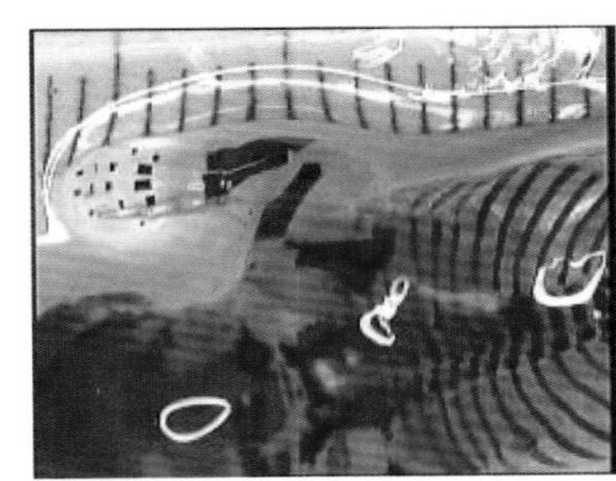

对照组
Control

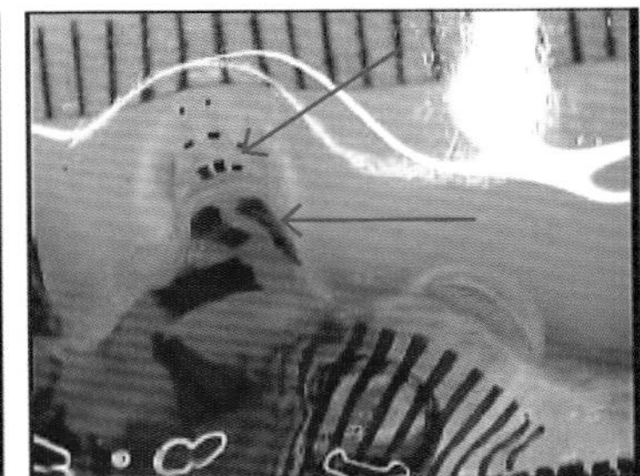

维生素 AD
Vitamin AD

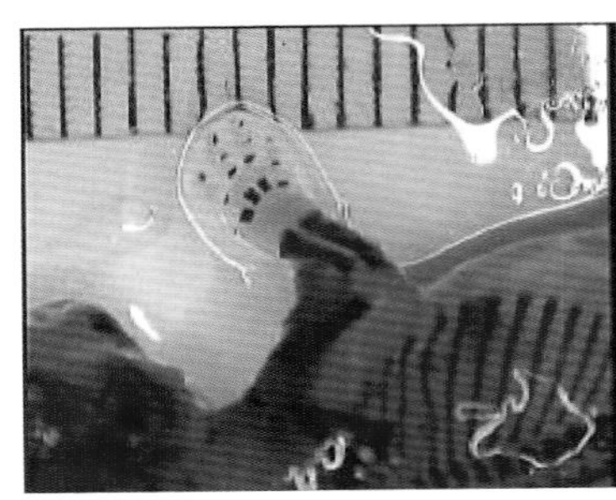

二月兰籽 -13
Orycho seeds-13

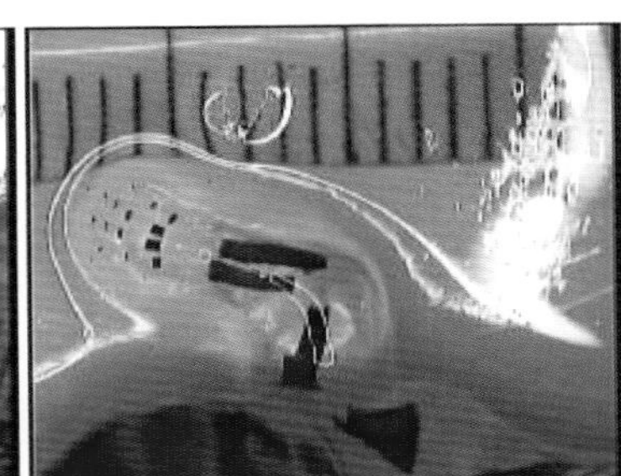

二月兰籽 -26
Orycho seeds-26

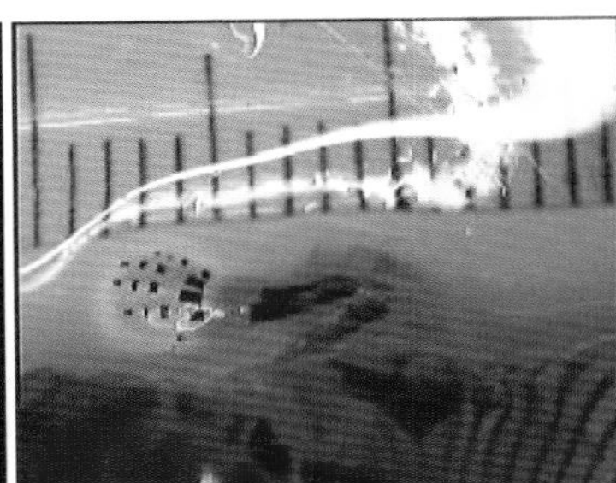

二月兰籽 -52
Orycho seeds-52

图 3-9 胎小鼠四肢骨和趾骨比较
Fig.3-9 Comparison of limb bones and digital bones of fetal mice

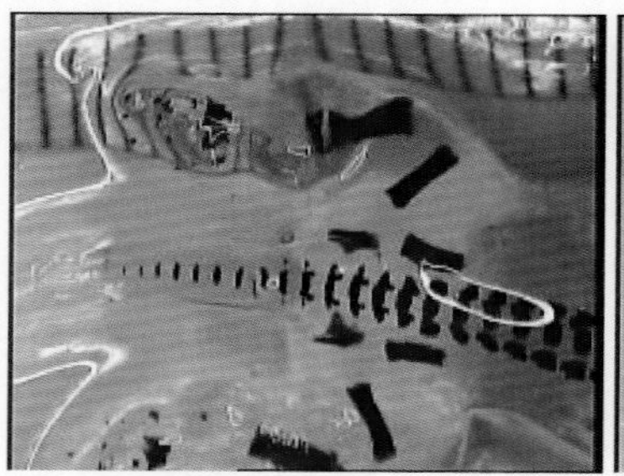

对照组
Control

维生素 AD
Vitamin AD

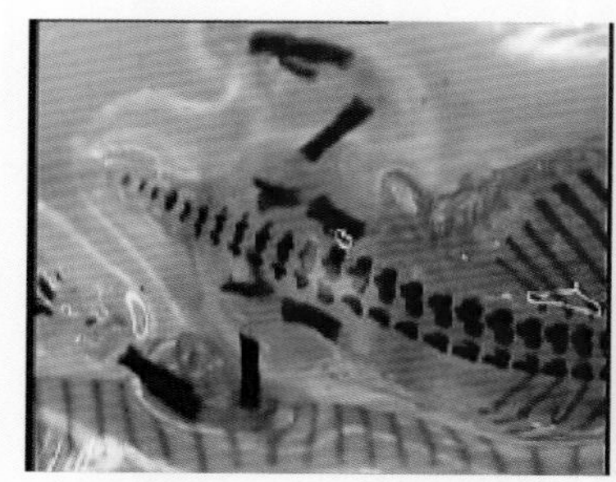

二月兰籽 -13
Orycho seeds-13

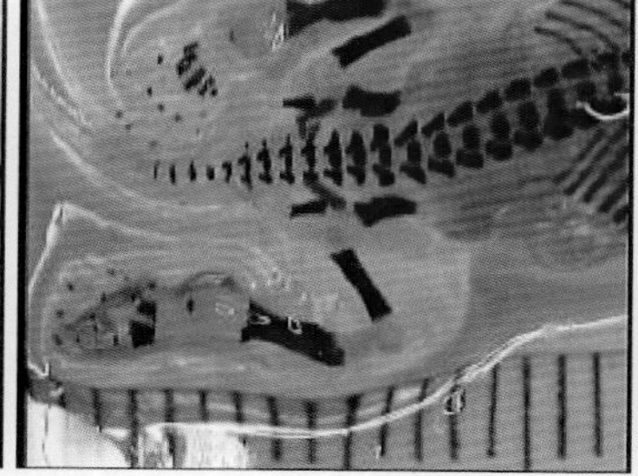

二月兰籽 -26
Orycho seeds-26

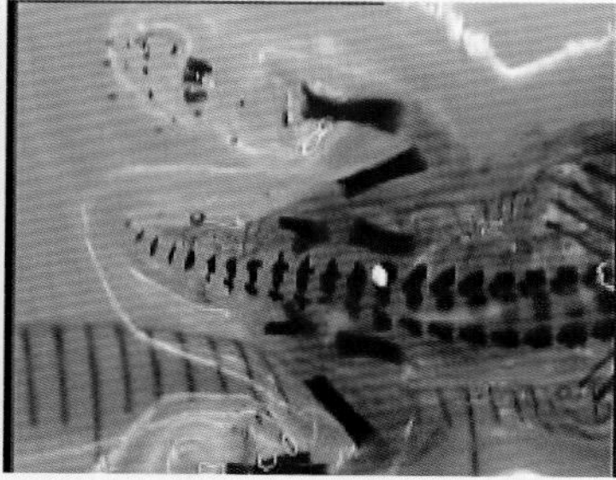

二月兰籽 -52
Orycho seeds-52

图 3-10 胎小鼠尾椎比较
Fig.3-10 Comparison of caudal vertebrae of fetal mice

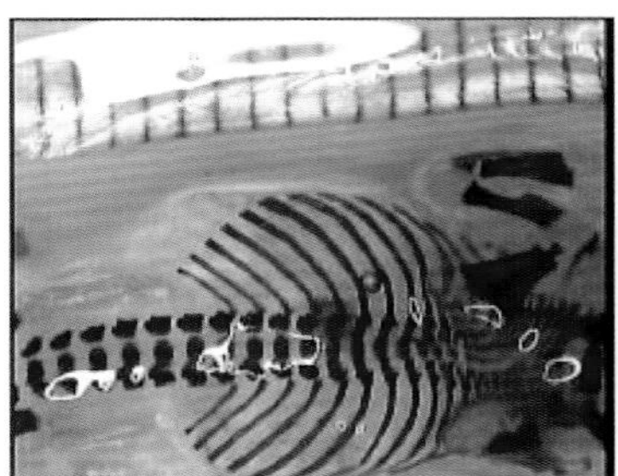

对照组
Control

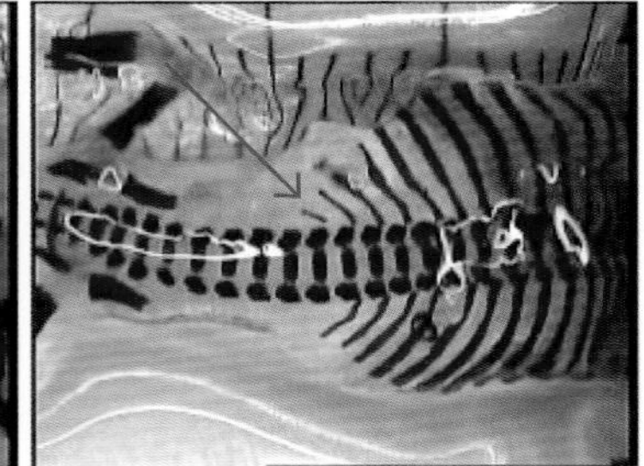

维生素 AD
Vitamin AD

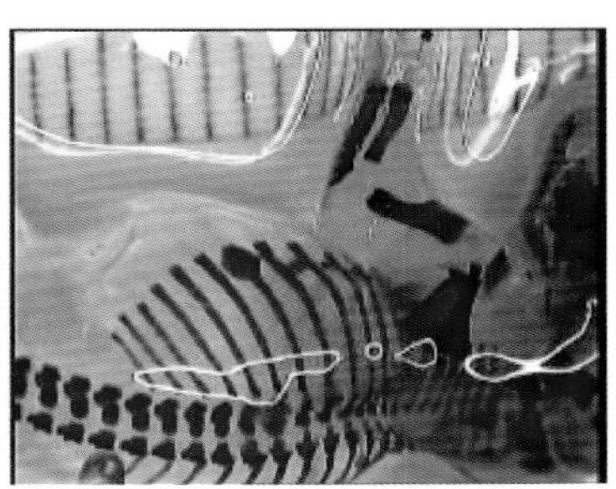

二月兰籽 -13
Orycho seeds-13

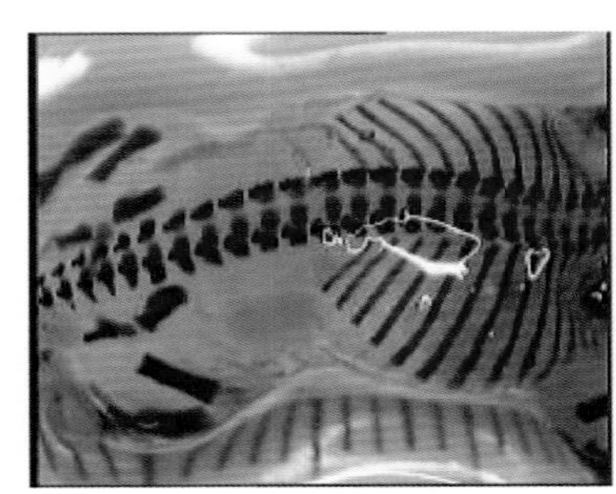

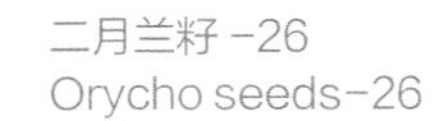

二月兰籽 -26
Orycho seeds-26

二月兰籽 -52
Orycho seeds-52

图 3-11 胎小鼠肋骨比较
Fig.3-11 Comparison of ribs of fetal mice

（6）胎小鼠内脏检查结果

胎小鼠内脏检查观察到维生素 AD 组胎鼠出现明显的唇裂、上腭裂、舌异常，其余各组未观察到类似现象（表3-18，图3-12和图3-13）。

(6) Results of visceral examination of fetal mice

Visceral examination of fetal mice showed that fetal mice in vitamin AD group had obvious cleft lip, upper cleft palate and abnormal tongue, but no similar phenomenon was observed in other groups (Table 3-18, Fig.3-12 and Fig.3-13).

表 3-18 胎小鼠内脏徒手切片检查出现畸形的百分比
Table 3-18 Percentage of malformation in visceral biopsy of fetal mice

分组 Group (g/kg)	腭、舌 Palate, tongue (%)	唇 Lip (%)	合计 Total (%)
对照组 Control	0.00	0.00	0.00
维生素 AD Vitamin AD	76.36	78.18	79.09
二月兰籽 -13 Orycho seeds-13	0.00	0.00	0.00
二月兰籽 -26 Orycho seeds-26	0.00	0.00	0.00
二月兰籽 -52 Orycho seeds-52	0.00	0.00	0.00

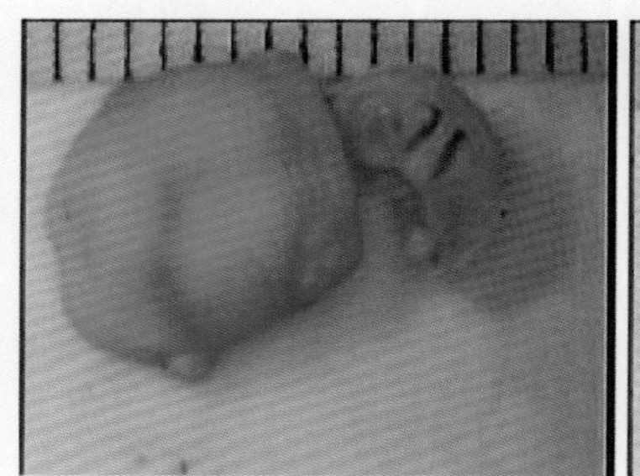

对照组
Control

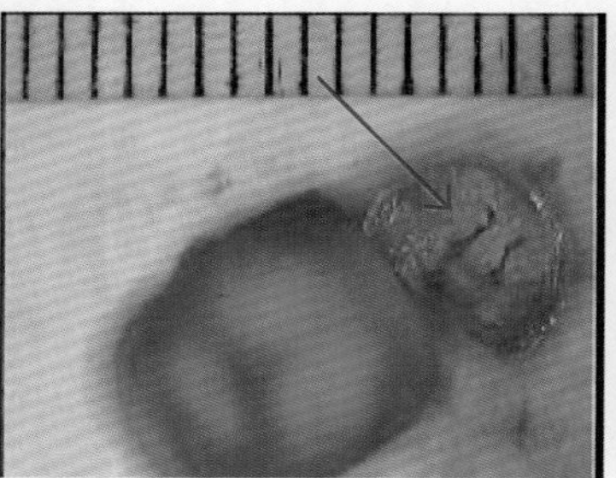

维生素 AD
Vitamin AD

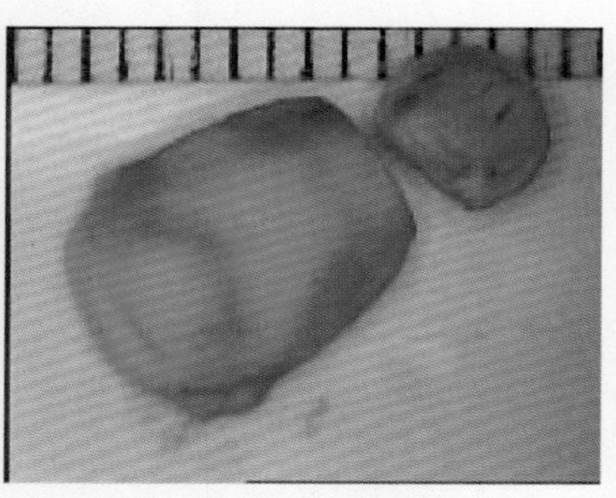

二月兰籽 -13
Orycho seeds-13

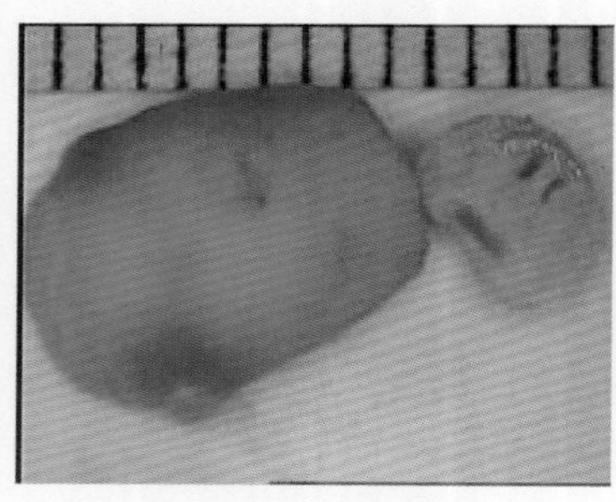

二月兰籽 -26
Orycho seeds-26

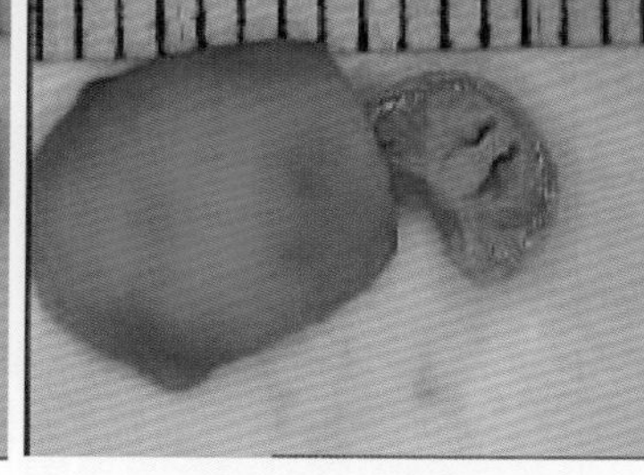

二月兰籽 -52
Orycho seeds-52

图 3-12 胎小鼠上腭比较
Fig.3-12 Comparison of the palate of fetal mice

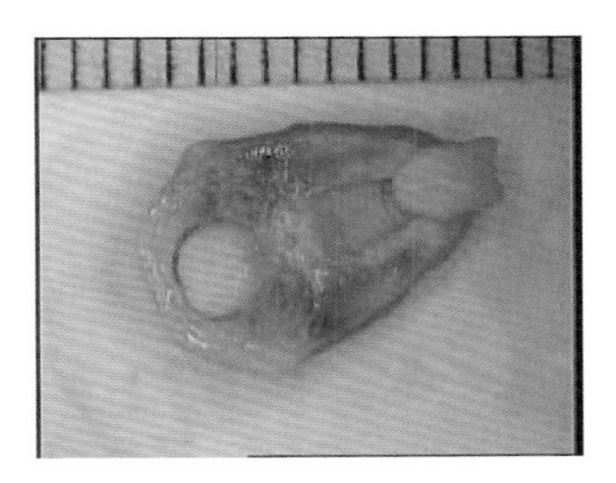

对照组
Control

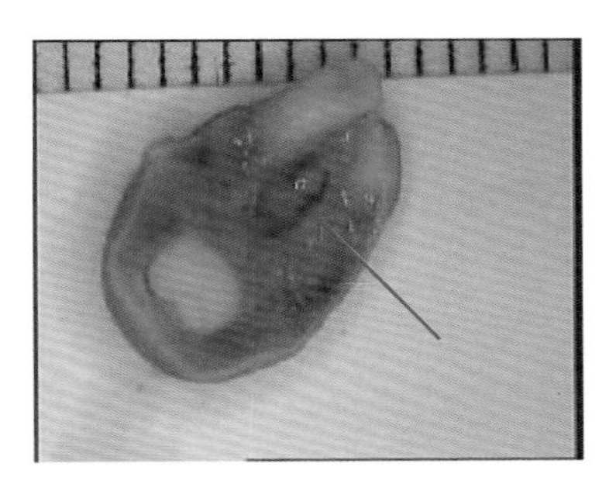

维生素 AD
Vitamin AD

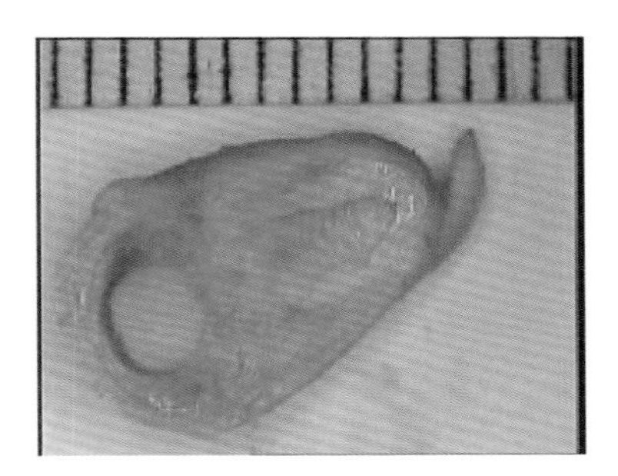

二月兰籽 -13
Orycho seeds-13

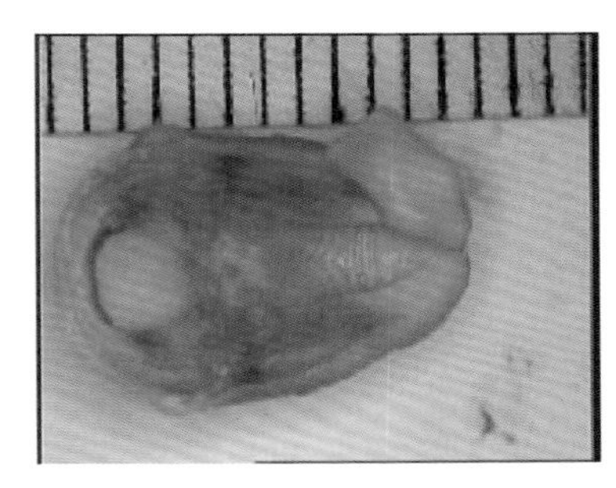

二月兰籽 -26
Orycho seeds-26

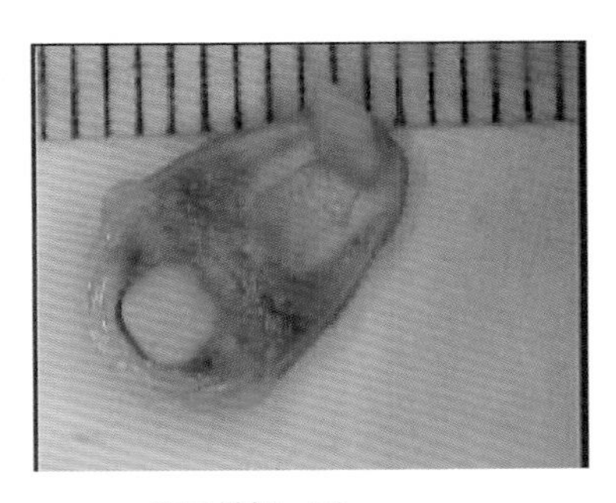

二月兰籽 -52
Orycho seeds-52

图 3-13 胎小鼠舌比较
Fig.3-13 Comparison of fetal mouse tongue

结论

致畸试验结果表明，二月兰籽对小鼠无致畸毒性。

Conclusion

Teratogenic test showed that orycho seeds had no teratogenic toxicity to mice.

第四章 二月兰保肝作用研究

Study on Liver-Protection Effect of orycho

二月兰抗毒品可卡因的肝毒性

背景

可卡因是一种非常容易上瘾的兴奋剂，由原产于南美洲的植物古柯叶（coca leaf）制成。

人们将可卡因作为毒品管控大概是 1903 年，这是可卡因社会管理的一个里程碑。这之前人们一直将其作为良性的兴奋剂来使用的。很多毒品都具有肝毒性。开发针对毒品中毒的解毒剂是毒理学的三大任务之一。

Anti-cocaine hepatotoxicity of orycho

Background

Cocaine is a very addictive stimulant, which is made from coca leaf, a plant native to South America.

It was about 1903 that people controlled cocaine as drug, which was a milestone in the social management of cocaine. Before that, people used it as a benign stimulant. Many drugs are hepatotoxic, Developing antidote for drug poisoning is one of the three major tasks of toxicology.

摘要

目的

研究二月兰籽水煎剂对可卡因致小鼠急性肝损伤的保护作用。

方法

50只雄性ICR小鼠随机分为对照组、模型组、二月兰籽水煎剂低、中、高剂量组，每组10只。二月兰籽水煎剂经口给予折合成生药量分别为21.5g/kg、43.0g/kg和86.0g/kg，对照组和模型组经口给予等量的蒸馏水，1次/天，连续4天；在实验的第4天，模型组、二月兰籽水煎剂 剂量组腹腔注射盐酸可卡因溶液(75mg/kg)，对照组腹腔注射等量的生理盐水。在染毒后24 h，所有动物眼眶内眦静脉取血，测定血清中谷丙转氨酶(GPT)、谷草转氨酶(GOT)和乳酸脱氢酶(LDH)的活性。留取肝组织测定丙二醛(MDA)和过氧化氢酶(CAT)的含量。留取部分肝组织做常规病理切片检查。

结果

与对照组比较，模型组小鼠血清GPT、GOT和LDH活性均显著上升，肝组织MDA含量升高，CAT含量明显下降。与模型组比较，低、中、高剂量的二月兰籽水煎剂组小鼠血清GPT、GOT和LDH活性均显著降低，并呈剂量依赖性。肝组织MDA含量明显下降，CAT含量显著升高。病理切片显示二月兰籽水煎剂能够明显减轻可卡因对肝组织的炎症性破坏。

结论

二月兰籽水煎剂对可卡因引起的急性肝损害有一定的保护作用。

Abstract

Objective

To study the protective effect of orycho seeds decoction on acute liver injury induced by cocaine in mice.

Methods

50 male ICR mice were randomly divided to control group, a model group, and low, medium and high dose groups of orycho seeds decoction, with 10 mice in each group. After oral administration, the orycho seeds decoction was converted into 21.5 g/kg, 43.0g/kg and 86. 0g/kg, the control group and the model group were given the same amount of distilled water orally, q.d for 4 consecutive days; On the fourth day of the experiment, the model group, cocaine hydrochloride solution (75mg/kg) was injected intraperitoneally in the dose group, and the same amount of normal saline was injected intraperitoneally in the control group. At 24h after exposure, the canthal veins in the orbit of all animals. The activities of glutamic pyruvic transaminase (GPT), glutamic oxaloacetic transaminase (GOT) and lactate dehydrogenase (LDH) in serum were measured. Determination of malondialdehyde (MDA) in liver tissue and the content of Catalase (CAT). Some liver tissues were taken for routine pathological examination.

Results

Compared with the control group, cocaine treated mice serum GPT, GOT and MDA content in liver tissue increased and CAT content decreased significantly. Compared with cocaine group, low, medium and high doses of orycho seeds decoction. The activities of GPT, GOT and LDH in serum of mice were significantly decreased in a dose-dependent manner. MDA content in liver tissue decreased significantly, while CAT content increased significantly. The histological section showed that the orycho seeds decoction could obviously reduce the inflammatory damage of cocaine to liver tissue.

Conclusion

The orycho seeds decoction has protective effect against cocaine- induced liver damage in mice.

除了二月兰以外，我们还开发了多个抗可卡因神经行为毒性和肝脏毒性的天然产物和合成化合物，形成了一个抗可卡因的物品组合。已经发表的抗可卡因物质论文目录如下，我们希望吸毒可卡因毒品的个体能够早日使用上这些物质来改善他们的健康。

1. 丁兆丰，贾凤兰，阮明，等．酮康唑对可卡因致小鼠肝毒性的保护作用．中国药理学与毒 理学杂志，2006，20(5):399-404.

2. 蔡青圆，陈虎彪，赵中振，等．五指毛桃拮抗毒品可卡因的肝毒性作用及其活性成分．中国中药杂志，2007，32（12）：1190-1193.

3. 姚青，高岭，贾凤兰，等．穿心莲内酯对可卡因致小鼠肝毒性的保护作用．宁夏医学杂志，2007，29（3）：208-209.

4. 邢国振，贾凤兰，阮明，等．大蒜素对可卡因致小鼠急性肝损伤的防治作用．中国药理学与毒理学杂志，2008：22（4）：284-290.

5. 姚青，张继荣，高岭，等．盐酸地尔硫䓬对小鼠可卡因肝损伤保护作用的研究．宁夏医学杂志，2010，32(12):1117-1119.

6. 吕艳，丁兆丰，马秋霞，等.1,3- 二苯 -1,3- 丙二酮对可卡因致小鼠肝毒性及神经毒性的保护作用．中国药物依赖性杂志，2011，20（2）：87-92.

7. 王璐，丁兆丰，贾凤兰，等．本米对可卡因所致小鼠急性肝损伤的保护作用，毒理学杂志，2011，25 (2): 93-96.

8. 吴学银，薛茹，刘昕，等．2,4- 二羟基二苯甲 酮对可卡因致小鼠肝毒性及神经毒性的保护作用．北京大学学报医学版，2012，44 (3): 421-425.

9. 魏鹏，刘伟霞，贾凤兰，等．8- 甲氧补骨脂素对可卡因致小鼠急性肝损伤的保护作用．中华中医药杂志，2013，28（3）：662-665.

10. 陈娟，王一超，崔蓉，等．1,3- 二苯 -1,3- 丙二酮对可卡因致小鼠神经递质含量变化的影响．北京大学学报（医学版），2016，48 (3): 398-402.

11. 王一超，孙艺，崔蓉，等．2,4- 二羟基二苯甲酮对可卡因致小鼠脑神经递质变化的影响．中国药物依赖性杂志，2016，26（4）：427-433.

In addition to orycho, our team has developed a number of natural products and synthetic compounds against neurobehavioral toxicity and hepatotoxicity of cocaine, forming an anti-cocaine combination. The published papers on anti-cocaine substances are listed as follows. We hope that individuals who take cocaine can use these substances as soon as possible to improve their health.

1.DING Zhaofeng, JIA Fenglan, RUAN Ming, et al. Protective effect of ketoconazole on cocaine-induced hepatotoxicity in mice. Chinese journal of pharmacology and toxicology, 2006，20(5):399-404.

2.CAI Qingyuan, CHEN Hubiao, ZHAO Zhongzhen, et al. Antagonistic effect of Wuzhi Maotao on the hepatotoxicity of cocaine and its active components, Chinese Journal of Traditional Chinese Medicine, 2007,32（12）：1190-1193.

3.YAO Qing, GAO ling, JIA Fenglan, et al. Protective effect of andrographolide on cocaine-induced hepatotoxicity in mice. Ningxia Journal of Medicine, 2007,29 (3): 208-209.

4.XING Guozhen, JIA Fenglan, RUAN Ming, et al. Preventive and therapeutic effects of allicin on acute liver injury induced by cocaine in mice. Chinese Journal of Pharmacology and Toxicology, 2008, 22（4）：284-290.

5.YAO Qing, ZHANG Jirong, GAO ling, et al. Protective effect of diltiazem hydrochloride on cocaine liver injury in mice. Ningxia Medical Journal, 2010,32(12):1117-1119.

6.LV Yan, DING Zhaofeng, MA Qiuxia, et al. Protective effect of 1,3- diphenyl -1,3- propanedione on hepatotoxicity and neurotoxicity induced by cocaine in mice. Chinese Journal of Drug Dependence, 2011, 20(2): 87-92.

7.WANG Lu, DING Zhaofeng, JIA Fenglan, et al. Protective effect of benmi on acute liver injury induced by cocaine in mice. Journal of Toxicology, 2011,25 (2): 93-96.

8.WU Xueyin, XUE Ru, LIU Xin, et al. Protective effect of 2,4- dihydroxybenzophenone on cocaine-induced hepatotoxicity and neurotoxicity in mice. Journal of Peking University Medical Edition, 2012, 44 (3): 421-425.

9.WEI Peng, LIU Weixia, JIA Fenglan, et al. Protective effect of 8- methoxypsoralen on acute liver injury induced by cocaine in mice. Chinese Journal of Traditional Chinese Medicine, 2013,28（3）：662-665.

10.CHEN Juan, WANG Yichao, CUI Rong, et al. Effect of 1,3- diphenyl -1,3- propanedione on changes of neurotransmitter content in mice induced by cocaine. Journal of Peking University (Medical Edition), 2016, 48 (3): 398-402.

11.WANG Yichao, SUN Yi, CUI Rong, et al. Effect of 2,4- Dihydroxybenzophenone on Changes of Brain Neurotransmitters Induced by Cocaine in Mice.Chinese Journal of Drug Dependence, 2016,26 (4): 427-433.

二月兰抗酒精的肝毒性

背景

酒精是臭名昭著的肝脏毒物，会导致各种病变，包括：

- ◆ 脂肪肝
- ◆ 酒精性肝炎
- ◆ 纤维化
- ◆ 肝硬化
- ◆ 肝癌

急需：目前还没有公认的酒精肝中毒的解毒剂。

Anti-alcoholic hepatotoxicity of orycho

Background

Drinking can lead to a variety of liver problems including:

- ◆ Steatosis
- ◆ Alcoholic hepatitis
- ◆ Fibrosis
- ◆ Cirrhosis
- ◆ Liver cancer

Urgent: There is no government-approved antidote to alcoholic hepatitic intoxication.

摘要

目的

通过建立小鼠急性酒精性肝损伤模型，探讨二月兰籽对急性酒精性肝损伤的保护作用。

方法

以二月兰籽为受试药物，用酒精制作小鼠化学性肝损伤模型。采用预防性给药的方式，二月兰籽低、中、高三个剂量组（9.0g/kg，18.0g/kg，36.0g/kg）经口灌胃。先给予受试药物30d，末次给药30min后，给予酒精染毒，16h后，测定血清中谷丙转氨酶（ALT）、甘油三酯（TG）、总胆固醇（TC）、低密度脂蛋白（LDL）、高密度脂蛋白（HDL）活性，留取肝脏组织石蜡包埋切片，HE染色，光镜下观察肝脏组织病理变化；制备肝匀浆，测定肝组织中还原型谷胱甘肽（GSH）、氧化型谷胱甘肽（GSSG）和丙二醛（MDA）含量，计算GSH/GSSG比值。

结果

酒精组血清ALT、LDL升高，血清TC、TG、HDL降低，肝组织MDA含量增加、GSH/GSSG比值降低，肝脏病理损伤明显，而二月兰籽剂量组明显缓解了血清指标的改变，肝组织MDA含量降低、GSH/GSSG比值增大，肝脏病理损伤明显改善。

结论

二月兰籽对酒精所致急性化学性肝损伤有保护作用。

Abstract

Objective

To explore the protective effect of orycho seed on acute alcoholic liver injury by an acute alcoholic mice liver injury model.

Methods

The mice liver injury model was made with alcohol. By preventive administration, three groups (9.0g/kg, 18.0g/kg, 36.0g/kg) of orycho seeds were given orally. After 30 days of administration, 30 minutes after the last administration, the drug was poisoned with alcohol. After 16 hours, the activities of alanine aminotransferase (ALT), triglyceride (TG), total cholesterol (TC), low density lipoprotein (LDL) and high density lipoprotein (HDL) in serum were measured. The paraffin- embedded sections of liver tissue were taken and stained with HE. The pathological changes of liver tissue were observed under light microscope. The liver homogenate was prepared, the contents of reduced glutathione (GSH), oxidized glutathione (GSSG) and malondialdehyde (MDA) in liver tissue were measured, and the GSH/GSSG ratio was calculated.

Results

In alcohol treated group, serum ALT and LDL increased, serum TC, TG and HDL decreased, liver MDA content increased, GSH/GSSG ratio decreased, and liver pathological damage was obvious. However, in orycho seed groups, the changes of serum indexes were alleviated, liver MDA content decreased, GSH/GSSG ratio increased, and liver pathological damage improved obviously.

Conclusion

Orycho seeds have protective effect on acute chemical liver injury induced by alcohol.

材料和方法

受试物

二月兰籽（产自北京），以 10 倍蒸馏水煎煮 3 次，每次 30min，滤液合并，过滤浓缩至所需量，备用。

金属硫蛋白基因敲除 MT(-/-) 小鼠，由日本国立环境研究所远山千春先生和佐藤雅彦先生提供，饲养于北京大学公共卫生学院毒理系专门清洁动物房，实验前 5 只一笼，饲养 3d，以适应环境。动物室和温度为 (23±2)℃，自动通风，明暗周期 12h/12h，自由摄食饮水。饲料为军事医学科学院实验动物中心配置的 SPF 饲料。

方法

参照酒精肝损伤模型的方法《保健食品检验与评价技术规范 (2003)》。

动物分组及处理

雄性 MT(-/-) 小鼠 50 只，20~22g，随机分为 5 组，分别是阴性组和酒精组，二月兰籽低、中、高三个剂量组（剂量分别为 9g/kg、18g/kg、36g/kg）。每日经口灌胃给予二月兰籽水煎剂，阴性组和酒精组给予蒸馏水，连续给药 30d。动物每天称重，按体重调整二月兰籽水煎剂的给药量。酒精组和二月兰籽各剂量组于试验结束时一次灌胃给予 75% 酒精（12mL/kg），阴性组给予蒸馏水，禁食 16h。

给予酒精后禁食 16h，然后经腹腔注射戊巴比妥钠溶液（60mg/kg）麻醉，腹主动脉采血，然后立即取出肝脏，用预冷的生理盐水洗净、吸干、称取肝脏湿重。留取肝左叶，用 10% 的福尔马林溶液固定，做常规病理学检查（石蜡包埋，HE 染色），光镜下观察肝脏形态学改变；另在肝脏左叶取一定量肝组织，匀浆后测定丙二醛（MDA）、还原性谷胱甘肽（GSH）以及氧化性谷胱甘肽（GSSG）含量，计算 GSH/GSSG 的比值。血液作生化检测，检测血液中谷丙转氨酶（ALT）、甘油三酯（TG）、总胆固醇（TC）、低密度脂蛋白（LDL）、高密度脂蛋白（HDL）。

Materials and methods

Subject matter

The orycho seeds (from Beijing) were decocted with 10 times distilled water for 3 times, each time for 30 min, the filtrates were combined, filtered and concentrated to the required amount for later use.

Male C57BL/6, MT(-/-) mice, kindly provided by Dr. Chiharu Thoyama and Dr. Masahiko Sato, National Institute of Environmental Research, Japan, were reared in the specialized clean animal room of Department of Toxicology, School of Public Health, Peking University. Before the experiment, 5 mice were reared in a cage for 3 days to adapt to the environment. The temperature in animal room and laboratory is (23±2)℃ , which is ventilated automatically. The light and dark period is 12h/12h, and people can eat and drink freely. The feed is SPF feed prepared by Experimental Animal Center of Academy of Military Medical Sciences.

Methods

Referring to the method of alcohol liver injury model (Ministry of Health of the People's Republic of China. Technical Specifications for Inspection and Evaluation of Health Food. Beijing: Ministry of Health of the People's Republic of China, 2003: 135-139).

Animal grouping and treatment

Fifty male C57BL/6, MT(-/-) mice, 20~22g, were randomly divided into five groups: negative group and alcohol group, and low, medium and high dose groups (the doses were 9g/kg, 18g/kg and 36g/kg, respectively). Every day, the orycho seeds decoction were given by oral gavage, and the negative group and alcohol group were given distilled water for 30 days. The animals were weighed every day, and the dosage of orycho seeds decoction was adjusted according to the weight. At the end of the experiment, 75% alcohol (12mL/kg) was given to the alcohol group and the orycho seed group, while the negative group was given distilled water and fasted for 16 hours.

After being given alcohol, mice fasted for 16 hours, then were anesthetized by intraperitoneal injection of 60mg/kg pentobarbital sodium solution, blood was collected from abdominal aorta, and then the liver was taken out immediately, washed with precooled normal saline, sucked dry, and the wet weight of the liver was measured. The left lobe of liver was fixed with 10% formalin solution, and routine pathological examination (paraffin embedding and HE staining) was performed to observe the morphological changes of liver under light microscope In addition, a certain amount of liver tissue was taken from the left lobe of liver, and the contents of malondialdehyde (MDA), reducing glutathione (GSH) and oxidizing glutathione (GSSG) were measured after homogenization, and the ratio of GSH/GSSG was calculated. Blood biochemical test, detection of alanine aminotransferase (ALT), triglyceride (TG), total cholesterol (TC), low density lipoprotein (LDL), high density lipoprotein (HDL).

血液生化指标测定

所取小鼠血液于室温中静置 30min 后，4000r/min 离心 10min，取上层血清备用。

采用 HITACHI-7020 型自动生化分析仪测定血清中谷丙转氨酶（ALT）、甘油三酯（TG）、总胆固醇（TC）、低密度脂蛋白（LDL）、高密度脂蛋白（HDL）的水平。

肝组织 MDA 含量测定

肝脏中 MDA 含量用 TBA 反应法检测。

肝组织谷肝甘肽测定

采用改良 Hisson 法测定肝组织中 GSH 和 GSSG 的含量。

数据处理

用 SPSS 18.0 统计学软件对本研究所得结果进行统计分析，数据用均数 ± 标准差，即 $\bar{x} \pm S$ 表示，用单因素方差分析进行处理。

Determination of blood biochemical indexes

After the collected mouse blood was allowed to stand at room temperature for 30min, it was centrifuged at 4000r/min for 10min to obtain the supernatant serum for later use.

The HITACHI-7020 automatic biochemical analyzer was used to determine the levels of serum alanine aminotransferase (ALT), triglyceride (TG), total cholesterol (TC), low density lipoprotein (LDL), and high density lipoprotein (HDL).

Determination of MDA content in live tissue

MDA content in the liver was detected by TBA reaction.

Determination of glutenin in live tissue

The modified Hisson method was used to determine the content of GSH and GSSG in liver tissue.

Statistical analysis

The results of this study were statistically analyzed by SPSS 18.0. The data were expressed by mean ± standard deviation, and were processed by one-way ANOVA.

结果

小鼠给予 75% 酒精 12mL/kg 后，在 16h 内酒精组有 2 只小鼠死亡，二月兰籽低剂量组有 2 只小鼠死亡，死亡小鼠解剖后胃呈饱食状，未见脏器明显异常。阴性组以及二月兰籽中剂量组和高剂量组没有出现动物死亡，详情见表 4-1 和表 4-2。

对急性酒精肝损伤小鼠血清 ALT、TC、TG 的影响

酒精组小鼠给予 75% 酒精 12mL/kg 后，小鼠血液中谷丙转氨酶（ALT）水平升高，血清总胆固醇（TC）、血清甘油三酯（TG）水平降低，与阴性组比较，差异具有统计学意义（$P<0.05$）。在二月兰籽剂量组中，随着二月兰籽剂量的增加，小鼠血清 ALT 水平逐渐降低，血清 TC、TG 水平逐渐升高。

对急性酒精肝损伤小鼠血清脂蛋白的影响

酒精组小鼠给予 75% 酒精 12mL/kg 后，酒精组小鼠血清高密度脂蛋白（HDL）水平降低、血清低密度脂蛋白（LDL）水平升高，与阴性组比较，差异有统计学意义（$P<0.05$）。在二月兰籽剂量组中，随着二月兰籽剂量的增加，小鼠血清 HDL 水平逐渐升高，血清 LDL 水平逐渐降低。二月兰籽高剂量组小鼠血清 HDL、LDL 与酒精组比较，差异均有统计学意义（$P<0.05$）。

对急性酒精肝损伤小鼠肝脏 MDA 的影响

酒精组小鼠给予 75% 酒精 12mL/kg 后，造成小鼠肝组织中丙二醛（MDA）水平的升高，与阴性组比较，差异具有统计学意义（$P<0.01$）。在二月兰籽剂量组中，随着二月兰籽剂量的增加，小鼠肝组织中 MDA 水平逐渐降低。

对急性酒精肝损伤小鼠肝脏谷胱甘肽的影响

酒精组小鼠给予 75% 酒精 12mL/kg 后，小鼠肝组织中 GSH/GSSG 比值降低，与阴性组比较，差异具有统计学意义（$P<0.05$）。在二月兰籽剂量组中，随着二月兰籽剂量的增加，小鼠肝组织中 GSH/GSSG 水平逐渐升高。

Results

After mice were treated with 75% alcohol at 12mL/kg , two mice in the alcohol group and two mice in the low-dose group died within 16h. The stomach of the dead mice was full after dissection, and no obvious organ abnormality was observed. There were no animal deaths in the negative, mid-and high-dose groups, as detailed in Table 4-1 and Table 4-2.

Effect on serum ALT, TC, TG of mice with acute alcohol liver injury

After mice in the alcohol group were treated with 75% alcohol at 12mL/kg, the levels of alanine aminotransferase (ALT) in the blood were increased, and the levels of serum total cholesterol (TC) and serum triglyceride (TG) were decreased, with a statistically significant difference as compared with those in the negative group (P<0.05). In the cymbidium goeringii seed dose group, with the increase of cymbidium goeringii seed dose, the serum ALT level of mice was gradually decreased, and the serum TC and TG levels were gradually increased.

Effect on serum lipoprotein of mice with acute alcohol liver injury

After mice in the alcohol group were treated with 75% alcohol at 12mL/kg, the serum high density lipoprotein (HDL) level of mice in the alcohol group was decreased and the serum low density lipoprotein (LDL) level was increased, and the difference was statistically significant as compared with that in the negative group (P<0.05). In the orycho seed dose group, with the increase of the orycho seed dose, the serum HDL level of the mice was gradually increased, and the serum LDL level was gradually decreased. The differences of serum HDL and LDL between the high dose group and alcohol group were statistically significant (P<0.05).

Effect on liver MDA of mice with acute alcohol-induced liver injury

After mice in the alcohol group were given 75% alcohol at 12mL/kg, the malondialdehyde (MDA) level in mouse liver tissue was increased, and the difference was statistically significant as compared with that in the negative group (P < 0.01). In the orycho seed dose group, MDA levels in mouse liver tissues were gradually decreased with the increase of the orycho seed dose.

Effect of glutathione on acute alcohol-induced liver injury in mice

After mice in the alcohol group were given 75% alcohol at 12mL/kg, the ratio of GSH/GSSG in mouse liver tissue was decreased, and the difference was statistically significant as compared with that in the negative group (P<0.05). In the orycho seed dose group, GSH/GSSG levels in the liver tissue of mice gradually increased with the increase of the orycho seed dose.

表 4-1 二月兰籽水煎剂对酒精小鼠致死和血清 ALT、TC、TG、HDL 和 LDL 的影响

Table 4-1 Effect of orycho seeds decoction on alcohol induced mouse lethal rate, serum ALT, TC , TG, HDL and LDL

分组 Groups（g/kg）	死亡数（总数）Dead (n)	ALT（IU/L）	TC（mmol/L）	TG（g/L）	HDL（mmol/L）	LDL（mmol/L）
对照 Control	0/10	61 ± 4	4.72 ± 1.54	5.47 ± 1.38	2.03 ± 0.16	0.20 ± 0.07
酒精 Alcohol	2/10	$129 \pm 69^{\#}$	$3.26 \pm 0.49^{\#}$	$2.38 \pm 1.27^{\#}$	$1.70 \pm 0.29^{\#}$	$0.32 \pm 0.08^{\#}$
二月兰籽 -9 Orycho seeds-9	2/10	$49 \pm 17^{*}$	3.33 ± 0.53	2.54 ± 1.28	$1.78 \pm 0.22^{*}$	0.27 ± 0.04
二月兰籽 -18 Orycho seeds-18	0/10	$46 \pm 6^{*}$	$3.61 \pm 0.14^{*}$	$4.53 \pm 1.44^{*}$	$1.80 \pm 0.07^{*}$	0.24 ± 0.04
二月兰籽 -36 Orycho seeds-36	0/10	$36 \pm 10^{*}$	$4.50 \pm 0.79^{**}$	$5.42 \pm 1.67^{**}$	$2.03 \pm 0.16^{*}$	$0.20 \pm 0.07^{*}$

注：数据为均数 ± 标准差。$^{\#}P < 0.05$（与对照组比较）；$^{*}P < 0.05$，$^{**}P < 0.01$（与酒精组比较）。

Data is expressed by mean ± SEM. $^{\#}P<0.05$ (compared with the control group); $^{*}P<0.05$, $^{**}P<0.01$ (compared with alcohol group).

表 4-2 二月兰籽水煎剂对酒精小鼠肝脏 MDA 和 GSH/GSSG 的影响
Table 4-2 Effects of orycho seeds decoction on liver MDA and GSH/GSSG of alcohol-treated mice

分组 Groups (g/kg)	MDA (mmol/g)	GSH/GSSG
对照组 Control	0.893±0.106	4.513±0.342
酒精 Alcohol	1.338±0.157##	3.739±0.498#
二月兰籽 -9 Orycho seeds-9	1.249±0.168	4.120±0.222*
二月兰籽 -18 Orycho seeds-18	1.144±0.130	4.428±0.463*
二月兰籽 -36 Orycho seeds-36	1.089±0.127*	4.644±0.959*

注：数据为均数 ± 标准差。#$P<0.05$（与对照组比较）；*$P<0.05$，**$P<0.01$（与酒精组比较）。
Data is expressed by mean ± SEM. #$P<0.05$ (compared with the control group); *$P<0.05$, **$P<0.01$ (compared with alcohol group).

图 4-1 二月兰籽对小鼠酒精肝病理表现的影响（HE 染色，×200）
A—对照组；B—酒精组；C—二月兰籽 -9 组；D—二月兰籽 -18 组；E—二月兰籽 -36 组
Fig.4-1 Effects of orycho seeds on the pathological manifestations of alcoholic hepatitis in mice (HE staining, ×200)
A—control group; B—alcohol group; C—orycho seeds -9 group; D—orycho seeds-18 group; E—orycho seeds-36 group

二月兰籽对急性酒精肝损伤小鼠肝脏病理影响

光镜下观察（HE 染色），阴性组小鼠肝脏肝小叶轮廓清楚，肝细胞胞质丰富，核大而圆，核仁清晰，细胞排列紧密（图 4-1A）。酒精组小鼠肝脏小叶轮廓不清晰，可以看到中央静脉周围细胞坏死，大量的炎性细胞浸润，同时可以看到细胞中出现弥漫性的圆形脂滴（图 4-1B）。二月兰籽低剂量组小鼠肝脏轮廓不清晰，中央静脉周围细胞坏死数量较酒精组有所减少，炎性细胞浸润变轻，只观察到中央静脉周围细胞中出现弥漫性的圆形脂滴（图 4-1C）。二月兰籽中剂量组小鼠肝脏轮廓变清晰，中央静脉周围只观察到少量的细胞坏死，炎性细胞浸润少见，中央静脉周围只有少量细胞出现圆形脂滴（图 4-1D）。二月兰籽高剂量组小鼠肝脏轮廓清晰，中央静脉周围只观察到个别细胞的坏死，未观察到炎性细胞，中央静脉周围细胞未观察到圆形脂滴（图 4-1E）。

Pathological effect of orycho seeds on live of mice with acute alcohol liver injury

Observation under a light microscope (HE staining) showed that the liver lobules of the mice in the negative group had a clear outline, and the hepatocytes were rich in cytoplasm, with large and round nuclei, clear nucleoli and tightly arranged cells (Fig.4-1A). In the alcohol group, the outline of the hepatic lobules was not clear, and cell necrosis around the central vein and a large amount of inflammatory cell infiltration could be seen, together with diffuse round lipid droplets in the cells (Fig.4-1B). In the low-dose group, the outline of the mouse liver was unclear, the number of cell necrosis around the central vein was reduced as compared with that in the alcohol group, and the inflammatory cell infiltration was milder. Only diffuse round lipid droplets in the cells around the central vein were observed (Fig.4-1C). In the medium-dose group, the liver contour became clear. Only a small amount of cell necrosis was observed around the central vein, and inflammatory cell infiltration was rare. Only a small number of cells around the central vein showed round lipid droplets (Fig.4-1D). The liver of the mice in the high dose group of orycho seeds was clearly delineated. Only individual cell necrosis and no inflammatory cells were observed around the central vein. No round lipid droplets were observed in the cells around the central vein (Fig.4-1E).

讨论

肝脏是进入体内的药物以及毒物代谢的主要场所，大多数的药物以及毒物在肝脏内通过生物转化而排出体外。ALT 在体内代谢过程中起着重要的作用，其活性与肝功能密切相关，TC 和 TG 的降低是急性重症肝炎的重要指标。HDL 的血液中密度最高、颗粒最小一种脂蛋白，是血脂代谢的基础物质，可以反映肝脏参与脂代谢能力。LDL 的主要功能是把胆固醇运输到全身各处细胞，运输到肝脏合成胆酸，血液中胆固醇升高以及肝脏受损时会引起 LDL 的升高。MDA 的含量常常可反映机体内脂质过氧化的程度，间接地反映出细胞损伤程度。GSH/GSSG 比值降低是脂质过氧化造成 GSH 耗竭的表现。因此，本研究采用血清中 ALT、TC、TG、HDL、LDL 以及肝脏中 MDA、GSH/GSSG 比值作为急性酒精肝损伤的指标。

酒精性肝病在我国近年来有逐渐增多的趋势，酗酒成为我国继病毒性肝炎之后导致肝损害的第二大病因。肝脏是酒精代谢、降解的主要场所，进入血液循环的乙醇约 90% 在肝脏内氧化。急性酒精性肝损伤的机制是当机体摄入大量乙醇后，在乙醇脱氢酶的催化下进行脱氢氧化，使三羧循环受阻和脂肪酸氧化减弱而影响脂肪的代谢，可致 α-磷酸甘油增多而促进 TG 合成，致使脂肪在肝细胞内沉积，因此脂质代谢异常为急性酒精性肝损伤的主要表现。

实验结果显示，二月兰籽可以明显降低急性酒精肝损伤导致小鼠血清中升高的 ALT、LDL 水平和肝组织中 MDA 含量，提高血清中 TC、TG、HDL 的水平和肝组织中 GSH/GSSG 比值。二月兰籽可能通过提高机体的抗氧化损伤能力，阻止 GSH 的耗竭，而阻断乙醇对肝细胞造成的过氧化损伤。病理结果显示，给予小鼠酒精后，造成酒精组小鼠肝脏的急性损伤，小鼠肝脏小叶轮廓不清晰，可以看到中央静脉周围细胞坏死，大量的炎性细胞浸润。二月兰籽各剂量组随着剂量的增加，肝脏病理损伤逐渐减轻，二月兰籽高剂量组小鼠肝脏病理情况已经基本接近阴性组，显示了二月兰籽可以有效抑制酒精对小鼠肝脏造成的急性损伤，对肝脏产生保护作用。

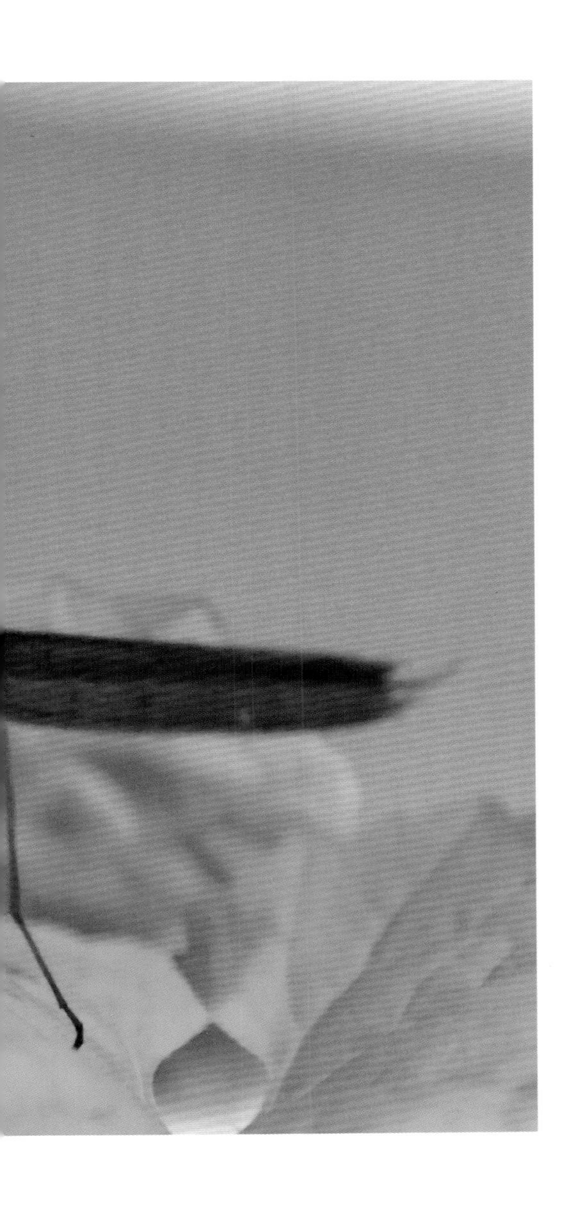

Discussion

The liver is the main organ for the metabolism of poisons entering the body. Most poisons are excreted from the body through biotransformation in the liver. ALT plays an important role in vivo metabolism, and its activity is closely related to liver function. The reduction of TC and TG are important indicators of acute severe hepatitis. HDL, the lipoprotein with the highest density and the smallest particle in the blood, is the basic substance for lipid metabolism, and it can reflect the ability of the liver to participate in lipid metabolism. The main function of LDL is to transport cholesterol to cells everywhere in the body and to the liver to synthesize cholic acid, which will cause the increase of LDL when cholesterol in the blood increases and the liver is damaged. The content of MDA can often reflect the degree of lipid peroxidation in the body, and indirectly reflect the degree of cell damage. The decrease in GSH/GSSG ratio was a manifestation of GSH depletion by lipid peroxidation. Therefore, in this study, ALT, TC, TG, HDL, LDL in serum and MDA, GSH/GSSG ratio in liver were used as indicators of acute alcohol-induced liver injury.

Alcoholic liver disease has been increasing in recent years in China, and alcohol abuse has become the second largest cause of liver damage in China after viral hepatitis. The liver is the main site for alcohol metabolism and degradation, and about 90% of the ethanol entering the blood circulation is oxidized in the liver. The mechanism of acute alcoholic liver injury is that after the body ingested a large amount of ethanol, it underwent dehydrogenation and oxidation under the catalysis of ethanol dehydrogenase, which blocked the tricarboxylic cycle and weakened fatty acid oxidation, thereby affecting fat metabolism, and inducing the increase of α-glycerophosphate to promote TG synthesis, as well as fat deposition in hepatocytes. Therefore, abnormal lipid metabolism is the main manifestation of acute alcoholic liver injury.

The experimental results showed that orycho seeds could significantly reduce the elevated ALT and LDL levels in serum and MDA content in liver tissue of mice induced by acute alcohol liver injury, and increase the levels of TC, TG and HDL in serum and the ratio of GSH/GSSG in liver tissue. The seeds of orycho may inhibit the depletion of GSH by improving the anti-oxidative damage ability of the body, and block the peroxidation damage to hepatocytes caused by alcohol. The pathological results showed that after the mice were given alcohol, they caused acute damage to the liver of mice in the alcohol group. The outline of lobules in the mouse liver was not clear, and cell necrosis around the central vein and a large number of inflammatory cell infiltration could be seen. The pathological damage of the liver was gradually alleviated with the increase of dose in each dosage group of orycho seed. The pathological condition of the liver of mice in the high dose group of orycho seed was almost close to that in the negative group, indicating that orycho could effectively inhibit the acute injury to the liver caused by alcohol and produce a protective effect on the liver.

结论

根据实验结果，可以得出氧化应激和脂类代谢障碍是乙醇诱导肝毒性发病机制的基础的部分。二月兰可以通过改善氧化应激和脂类代谢来抑制乙醇诱导的肝毒性。

Conclusion

According to the experimental results, it can be concluded that oxidative stress and lipid metabolism disorder are the basic parts of the pathogenesis of ethanol-induced hepatotoxicity. Orycho can inhibit ethanol-induced hepatotoxicity by improving oxidative stress and lipid metabolism.

二月兰抗四氯化碳的肝毒性

背景

我们发现二月兰籽有明显的肝脏保护作用后，考虑到二月兰是叶类蔬菜，所以我们对二月兰全草的肝脏保护作用进行了检验。使用的肝脏毒性模型是经典的四氯化碳（CCl_4）模型，这也是保健食品中保肝功能检测中的规定项目。

Anti-Carbon Tetrachloride - hepatotoxicity of orycho

Background

After we found that the seeds of orycho have obvious hepatoprotective effect, considering that orycho is a leafy vegetable, we have examined the hepatoprotective effect of the whole plant of orycho. The hepatotoxicity model used is the classic carbon tetrachloride (CCl_4) model, which is also a prescribed item in liver protective function test for health food.

摘要

目的

评价二月兰全草对四氯化碳肝损伤的保护作用。

方法

采用 ICR 小鼠，以四氯化碳作为模型药物制造肝损伤模型。将小鼠随机分为溶剂对照组、模型组、二月兰水煎剂低、中、高剂量组（小鼠给予的生药量分别为 21.5g/kg、43.0g/kg 和 86.0g/kg）。经口给受试物，剂量为 0.02L/kg，对照组和模型组经口给予等量的生理盐水，1 次 / 天，连续 4 天；在实验的第 4 天，模型组、二月兰各组以四氯化碳染毒。在染毒后 24h，所有动物取血，检测血清中 ALT、AST 和 LDH 的含量。解剖取肝脏，留取肝组织匀浆测定组织中 MDA 的含量。取肝左叶进行常规病理切片，HE 染色，光镜下观察组织病理学变化。

结果

与对照组比较，模型组小鼠血清中 ALT、AST 和 LDH 活力明显升高，肝组织 MDA 含量上升，肝脏组织出现明显的肝细胞变性坏死。二月兰水煎剂组的小鼠血清中 ALT、AST 和 LDH 活力与模型组相比明显降低，肝脏组织病理损伤明显减轻，肝组织 MDA 含量降低。

结论

二月兰全草对四氯化碳所致急性化学性肝损伤有保护作用。

Abstract

Objective

To evaluate the protective effect of orycho whole plant on carbon tetrachloride-induced liver injury.

Methods

The liver injury model was established in ICR mice using carbon tetrachloride as the model drug. The mice were randomly divided into the solvent control group, model group, and the low, medium and high dose groups of orycho decoction. The crude orycho doses given to the mice were 21.5g/kg, 43.0g/kg and 86.0g/kg, respectively. The subjects were orally administrated with the dose of 0.02L/kg, and the control group and the model group were orally administrated with the same amount of normal saline, once a day, for four successive days. On the fourth day of the experiment, the model group and the orycho groups were administered with carbon tetrachloride. At 24h after administration, blood was taken from all animals and the serum ALT, AST and LDH contents were measured. The liver was dissected and the liver tissue homogenate was retained for determination of MDA content in the tissue. The left lobe of the liver was taken for routine pathological sectioning, HE staining and histopathological observation under the light microscope.

Results

Compared with the control group, the activities of ALT, AST and LDH in serum of mice in the model group were significantly increased, the MDA content in liver tissue was increased, and obvious degeneration and necrosis of hepatocytes were observed in the liver tissue. Compared with the model group, the activities of ALT, AST and LDH in serum of mice in the orycho decoction group were significantly reduced, the pathological damage in liver tissue was significantly alleviated, and the MDA content in liver tissue was reduced.

Conclusion

Orycho whole plant has protective effect on acute chemical liver injury induced by carbon tetrachloride.

材料和方法

受试物

二月兰全草，产自北京，干燥后保存使用。以 10 倍蒸馏水煎煮 3 次，每次 30min，滤液合并，过滤浓缩至所需量的水煎剂，备用。

动物分组及处理

将 60 只 ICR 小鼠随机分为溶剂对照组、CCl_4 组、二月兰水煎剂低、中、高剂量组（折合干燥二月兰分别为 21.5g/kg、43.0g/kg 和 86.0g/kg）和单独二月兰组（86.0g/kg），每组 10 只。水煎剂给予剂量为 0.02L/kg，对照组和模型组经口给予等量的蒸馏水，1 次 / 天，连续 4 天；在实验的第 4 天，CCl_4 组、二月兰 3 个剂量组腹腔注射 CCl_4 玉米油溶液（2%）5mL/kg，对照组腹腔注射等量的玉米油。在染毒后 24h，所有动物眼眶内眦静脉取血，制备血清，检测 ALT、AST 和 LDH 的活性。脱臼处死小鼠，立即解剖取肝脏，称重。留取肝组织匀浆测定组织 MDA 的 含量。将肝左叶浸于质量分数为 10%甲醛溶液中固定，进行常规病理切片，HE 染色，光镜下观察小鼠肝脏的组织病理学变化。

血液生化指标测定

采用 HITACHI-7020 型自动生化分析仪测定血清中谷丙转氨酶（ALT）、甘油三酯（TG）、总胆固醇（TC）、低密度脂蛋白（LDL）、高密度脂蛋白（HDL）的 水平。

肝组织 MDA 含量测定

肝脏中 MDA 含量用 TBA 反应法检测（同酒精肝试验）。

数据处理

用SPSS 18.0统计学软件对本研究所得结果进行统计分析，数据用均数 ± 标准差表示，用单因素方差分析进行处理。

Materials and methods

Materials

The whole plant of orycho, collected in Beijing, is stored and used after drying. Decocting with 10 times distilled water for 3 times, each time for 30min, mixing filtrates, filtering and concentrating to obtain the required amount of decoction for later use.

Animal grouping and treatment

Sixty ICR mice were randomly divided into solvent control group, CCl_4 group, low, medium and high dose groups (21.5g/kg, 43.0g/kg and 86.0g/kg, respectively), orycho control group (86.0g/kg), with 10 mice in each group. The dose of orycho decoction was 0.02L/kg, and the control group and model group were given the same amount of distilled water orally, once a day for 4 days. On the 4th day of the experiment, CCl_4 group and the three dosage groups were intraperitoneally injected with CCl_4 corn oil solution (2%) 5mL/kg, while the control group was intraperitoneally injected with the same amount of corn oil. Twenty-four hours after exposure, blood was taken from the canthus vein of all animals to prepare serum and detect the activities of ALT, AST and LDH. The mice were killed after dislocation, and the livers were dissected immediately and weighed. The content of MDA in liver homogenate was determined. The left lobe of liver was immersed in 10% formaldehyde solution for fixation, and the histopathological changes of mouse liver were observed by routine pathological section and HE staining under light microscope.

Determination of blood biochemical indexes

The levels of alanine aminotransferase (ALT), triglyceride (TG), total cholesterol (TC), low density lipoprotein (LDL) and high density lipoprotein (HDL) in serum were measured by HITACHI- 7020 automatic biochemical analyzer.

Determination of MDA content in liver tissue

MDA content in liver was detected by TBA reaction method.

Data processing

The results of this study were statistically analyzed by SPSS 18.0. The data were expressed by mean standard deviation, and were processed by one-way ANOVA.

结果

小鼠血清中 ALT、AST 和 LDH 活力的变化

CCl_4 组小鼠血清 ALT、AST 和 LDH 活力明显升高。随着二月兰剂量的增高，ALT 下降的程度增大，并且高剂量组已经接近正常水平。单独二月兰组，小鼠血清中 ALT 活性没有明显的改变（表 4-3）。

Results

Changes of ALT, AST and LDH activities in mouse serum

The activities of ALT, AST and LDH in serum of mice in CCl_4 group were significantly increased. With the increase of the dose of orycho, the degree of ALT decrease increased, and the high dose group was close to the normal level. There was no obvious change in serum ALT activity of mice in the orycho control group alone (Table 4-3).

表 4-3 二月兰抗 CCl_4 所致小鼠肝功酶的变化
Table 4-3 Effect of orycho on CCl_4 induced changes of liver functions

分组 Groups (g/kg)	ALT (IU/L)	AST (IU/L)	LDH (IU/L)
对照 Control	77±22	125±32	1699±208
CCl_4	2035±1137##	706±389#	3981±1528#
二月兰 -21.5 Orycho -21.5	1179±1787	469±564	3079±2052
二月兰 -43.0 Orycho -43.0	205±255**	254±216*	1962±596*
二月兰 -86.0 Orycho -86.0	78±21**	118±56*	1622±400*
二月兰对照 Orycho control	63±12*	—	—

注：数据为均数 ± 标准差。#P < 0.05（与对照组比较），##P < 0.01（与对照组比较）；*P < 0.05，**P < 0.01（和模型组比较）。
Data is expressed by mean ±SEM. #P<0.05 (compared with the control group); *P<0.05, **P<0.01 (compared with CCl_4 group).

表 4-4 小鼠肝 MDA 的变化
Table 4-4 Changes of mouse liver MDA

分组 Groups（g/kg）	MDA（μmol/g）
对照 Control	0.192±0.109
CCl_4	0.423±0.120#
二月兰 -21.5 Orycho -21.5	0.212±0.037*
二月兰 -43.0 Orycho -43.0	0.313±0.012
二月兰 -86.0 Orycho -86.0	0.284±0.101
二月兰对照 Orycho control	0.167±0.050

注：数据为均数 ± 标准差。#$P<0.05$（与对照组比较）；*$P<0.05$（和模型组比较）。
Data is expressed by mean ± SEM. #$P<0.05$ (compared with the control group); *$P<0.05$, **$P<0.01$ (compared with CCl_4 group).

小鼠肝 MDA 的变化

与溶剂对照组比较，CCl_4 组小鼠肝脏组织中 MDA 的含量明显升高（$P<0.05$）。二月兰能够抵抗这种上升（表 4-4）。

Changes of mouse liver MDA

Changes of MDA in mouse liver tissue Compared with the solvent control group, the content of MDA in liver tissue of mice in CCl_4 group was significantly higher ($P<0.05$). Orycho can resist this rise (Table 4-4).

肝组织病理变化

对照组小鼠肝脏表面光滑，色泽暗红，质地柔软。光学显微镜下可见肝小叶轮廓清晰，肝细胞胞质丰富，含嗜碱性物质较多，核大而圆，核仁清晰，肝细胞索围绕中央静脉呈放射状排列（图4-2A）；CCl_4 组小鼠肝脏肿大，质地较脆，表面粗糙有很多小的灰黄色点状坏死灶，个别肝脏可见明显小片状淤血，在光学显微镜下以广泛存在的肝脏小叶中央静脉周围的坏死为主，坏死细胞轮廓不清，胞核固缩或已溶解破碎，肝细胞索的正常形态被破坏，中间带可见细胞空泡样变及炎症细胞浸润（图4-2B）；与 CCl_4 组相比，随着二月兰剂量的增加，肝细胞变性、肿大等病理损伤逐渐减轻，肝细胞索形态逐渐恢复（图4-2C~图4-2E）。二月兰对照组的肝脏大体和病理切片的表现均与对照组无明显改变（图4-2F）。

Pathological changes of liver tissue

The liver of mice in control group has smooth surface, dark red color and soft texture. Under the light microscope, it can be seen that the outline of liver lobules is clear, the cytoplasm of liver cells is rich, there are more basophils, the nucleus is large and round, the nucleolus is clear, and the liver cell cords are radially arranged around the central vein (Fig.4-2A); In CCl_4 group, the liver of mice is swollen, the texture is brittle, the surface is rough, and there are many small gray-yellow spot necrosis foci. Some livers can be seen with obvious platelet congestion. Under the optical microscope, the necrosis around the central vein of liver lobule is widespread, the outline of necrotic cells is unclear, the nucleus is contracted or dissolved and broken, the normal shape of liver cell cord is destroyed, and cell vacuolation and inflammatory cell infiltration can be seen in the middle zone (Fig.4-2B); Compared with CCl_4 group, with the increase of the dose of orycho, the pathological damage such as liver cell degeneration and swelling gradually decreased, and the shape of liver cell cord gradually recovered (Fig.4-2C~Fig.4-2E). There was no change in the liver gross and pathological sections of the orycho control group (Fig.4-2F).

图 4-2 各组小鼠肝脏组织病理变化（HE，×200）

A—对照组，肝脏结构正常；B-CCl_4 组，肝脏小叶中央静脉周围出现坏死，肝细胞索正常形态被破坏，中间带可见细胞空泡样变及炎症细胞浸润；C、D、E—二月兰低、中、高剂量组，肝脏病理改变明显改善；F—二月兰对照组，肝脏与对照组无明显变化

Fig.4-2 Pathological changes of liver tissue of mice in each group (He,×200)

A—The liver structure of control group is normal; B—CCl_4 group, necrosis occurred around the central vein of hepatic lobule, the normal shape of hepatic cord was destroyed, and cell vacuolation and inflammatory cell infiltration were seen in the middle zone C, D, E—The pathological changes of liver in the low, medium and high dose groups were obviously improved. F—orycho alone control, there was no changes

用 DXM1200F 型摄像系统拍照后用 Photoshop7.0 图像处理软件对病理图片坏死和变性组织面积所占比例进行半定量分析。对照组无病理损伤。二月兰剂量组能够明显减小肝脏损伤的面积，并且剂量越大，损伤的面积越小（表 4-5）。

After taking pictures with DXM1200F camera system, the proportion of necrotic and degenerated tissue area in pathological pictures was semi-quantitatively analyzed with Photoshop7.0 image processing software. There was no pathological injury in the control group. The area of liver injury can be obviously reduced in the dose group, and the larger the dose, the smaller the area of liver injury (Table 4-5).

表 4-5 小鼠肝病变面积变化

Table 4-5 Changes of liver damage area

分组 Groups（g/kg）	损伤面积 Damage area（%）
对照 Control	0±0
CCl_4	41.6±13.7##
二月兰 -21.5 Orycho -21.5	31.5±4.5*
二月兰 -43.0 Orycho -43.0	17.4±3.2**
二月兰 -86.0 Orycho -86.0	7.4±2.8**

注：数据为均数 ± 标准差。经 t 检验，##P< 0.01（与对照组比较）；*P<0.05, **P<0.01（与模型组比较）。

Data is expressed by mean ±SEM. Compared with the model group, *P<0.05, **P<0.01 by t test.

讨论

CCl_4 进入体内在细胞色素 P-450 催化下经肝细胞微粒体 NADPH 依赖的电子传递系统代谢为三氯甲基自由基（$\cdot CCl_3$）和过氧化三氯甲基自由基（$\cdot OOCCl_3$），这些自由基可与不饱和 脂肪酸共价结合，使其脱氢而生成脂质自由基（L·）和过氧化脂质自由基（LOO·），致使脂质过氧化增强，进而引起细胞膜的破坏、酶的失活和毛细血管通透性增高等，最终导致肝细胞损伤坏死。小鼠 CCl_4 急性中毒时，染毒 24h 后血清中肝酶达到峰值，12～48h 肝脏均呈现典型的 CCl_4 中毒的改变，主要表现为肝小叶中央静脉周围的肝细胞坏死以及气球样变。

实验结果显示，CCl_4 小鼠肝脏组织中 MDA 的含量明显增高。但是在预防性给予二月兰后，MDA 的增高受到抑制。这说明中二月兰有保护肝脏细胞膜，抗脂质过氧化的作用。

在病理方面，二月兰水煎剂能够明显改善小鼠的肝脏的大体表现和病理变化。低剂量组即出现肝脏状况的改善，随着剂量的增加这种趋势变得更加明显，并接近对照组小鼠。病理损伤的面积也有所减少。这进一步证明了二月兰对于化学性肝损伤的保护作用。

Discussion

CCl_4 enters the body and is metabolized into trichloromethyl radical ($\cdot CCl_3$) and trichloromethyl peroxide radical ($\cdot OOCCl_3$) by NADPH-dependent electron transport system of hepatocyte microsomes under the catalysis of cytochrome P-450. These radicals can covalently bind with unsaturated fatty acids and dehydrogenate them to generate lipid free radicals (L·) and lipid peroxide free radicals (LOO·), thus enhancing lipid peroxidation, causing cell membrane destruction, enzyme inactivation and capillary permeability. During acute CCl_4 poisoning in mice, the liver enzymes in serum reached the peak after exposure for 24 hours, and the liver showed typical CCl_4 poisoning changes during 12~48 hours, mainly manifested as hepatocyte necrosis and balloon-like changes around the central vein of hepatic lobules. The experimental results showed that the content of MDA in liver tissue of CCl_4 mice increased significantly. However, the increase of MDA was inhibited after the preventive administration of Orchidaceae. This shows that the Chinese cymbidium can protect the liver cell membrane and resist lipid peroxidation.

In the pathological aspect, the orycho decoction can obviously improve the gross performance and pathological changes of mouse liver. The liver condition improved in low dose group, and this trend became more obvious with the increase of dose, which was close to the control group mice. The area of pathological injury also decreased. This further proves the protective effect of orycho on chemical liver injury.

结论

根据实验结果，可以得出二月兰可以通过改善氧化应激来抑制四氯化碳诱导的肝毒性。

Conclusion

According to the experimental results, it can be concluded that orycho can inhibit the hepatotoxicity induced by carbon tetrachloride by improving oxidative stress.

二月兰抗肝硬化和脂肪肝活性

背景

肝硬化是由多种形式的肝病和病症引起的肝脏瘢痕（纤维化）的晚期阶段，例如病毒性肝炎和酒精中毒等慢性中毒性肝损伤。现在还没有具体的治疗方法可以治愈肝硬化。寻找治疗肝硬化的药物是毒理学中最具挑战性的任务。

Anti-cirrhosis and anti-fat liver activity of orycho

Background

Cirrhosis is a late stage of scarring (fibrosis) of the liver caused by many forms of liver diseases and conditions, such as virus hepatitis and chronic toxic liver injury such as alcoholism. There are not specific treatments that can cure cirrhosis. Finding drugs to treat liver cirrhosis is the most challenging task in toxicology.

摘要

目的

建立四氯化碳（CCl_4）小鼠肝纤维化模型，研究二月兰籽水煎剂对 CCl_4 所致小鼠肝纤维化和脂肪变的保护作用。

方法

采用 CCl_4 制备小鼠肝纤维化模型，将 ICR 雄性小鼠随机分为 6 组：阴性对照组、CCl_4 模型组、二月兰籽水煎剂低、中、高剂量组（9g/kg，18g/kg，36g/kg）和二月兰籽水煎剂单独组（36g/kg）。实验开始前 1 周二月兰籽水煎剂各剂量组和单独组小鼠每天灌胃给予相应剂量的水煎剂，阴性对照组和 CCl_4 模型组给予等体积蒸馏水。实验第 1 周开始，每周三和周四在二月兰各剂量组给予二月兰籽水煎剂 1h 后，CCl_4 模型组和二月兰籽各剂量组背部皮下注射 0.1mL/kg 的 CCl_4，阴性对照组和二月兰单独组给予等体积的橄榄油，连续 16 周。末次 CCl_4 染毒 24h 后，麻醉后内眦静脉取血测血清丙氨酸氨基转移酶（ALT）、天冬氨酸氨基转移酶（AST）、乳酸脱氢酶（LDH）、碱性磷酸酶（ALP）、总胆红素（T-BIL）作为肝功能 的指标；以总蛋白（TP）、白蛋白（ALB）及白球比（A/G）作为肝脏合成蛋白的指标；以血尿素氮（BUN）和血肌酐（Cr）作为肾损伤的指标；以总胆固醇（TC）、低密度脂蛋白（LDL）、高密度脂蛋白（HDL）作为肝脏脂肪变性的指标；以羟脯氨酸（HYP）联合肝脏蛋白合成功能作为纤维化的指标；以肝组织中丙二醛（MDA）、还原型谷胱甘肽（GSH）和氧化型谷胱甘肽（GSSG）含量及其比值（GSH/GSSG）作为肝组织脂质过氧化和氧化应激的指标。同时留取肝、肾组织进行苏木精－伊红染色法（HE 染色），Van Gieson（V.G）染色法和脂肪染色法进行病理检查，并进行半定量计算。

结果

与阴性对照组相比，CCl_4 模型组小鼠血清中 ALT、AST、LDH、T-BIL、ALP、HBDH、ALB 和 CREA 水平明显升高，TC、HDL、LDL 和 BUN 水平下降，A/G 比值升高，TP 和 TG 水平 变化不大；肝组织中 HYP 和 MDA 水平明显升高，GSH/GSSG 比值明显降低。与 CCl_4 模型组相比较，二月兰籽水煎剂能明显降低血清 ALT、AST 和 LDH 水平，使 T-BIL、ALP、HBDH、ALB 和 CREA 水平有所下降，明显升高血清中 TC 水平，使 HDL 和 LDL 水平稍有升高；明显降低肝组织中 HYP 含量，使肝组织中 MDA 水平降低。病理组织学检查显示二月兰籽对 CCl_4 所致肝组织胶原纤维化和脂肪变具有良好的改善作用。

结论

二月兰籽水煎剂对 CCl_4 所致小鼠肝硬化和脂肪肝具有较好的保护作用。

Abstract

Object

To investigate the protective effect of orycho on CCl_4 induced liver fibrosisin mice.

Method

CCl_4 were experimented as the liver fibrosis model. The ICR male mice were randomly divided into 6 groups: negative control group, CCl_4 model group, orycho groups(9g/kg, 18g/kg and 36g/kg) and orycho single group (36g/kg). Each orycho seed dose group mice were given the same dose orycho seed by gastric gavage every day 1 week before the beginning of the experiment, and the negative control group and CCl_4 model group were given the same volume of distilled water. From the 1st week, 1h after given the aqueous extract of orycho seed, the CCl_4 model group and orycho groups (9g/kg, 18g/kg and 36g/kg) were subcutaneous injected with CCl_4 at the dose of 0.1mL/kg every Wednesday and Thursday for continuous 16 weeks, and the negative control group and orycho single group (36g/kg) were given the same volume of olive oil. 24h after the last injection of CCl_4, serum aminotransferase (ALT), aspartate aminotransferase (AST), lactate dehydrogenase (LDH), kaline phosphatase (ALP), total bilirubin (T-BIL), total protein (TP), albumin (ALB), A/G ratio, blood urea nitrogen (BUN), Creatinine (CREA), total cholesterol (TC), low density lipoprotein (LDL), high density lipoprotein (HDL) were measured to evaluate the function of liver as well as hydroxyproline (HYP), malondialdehyde (MDA), reduced glutathione (GSH) and Oxidized glutathione (GSSG) in liver and GSH/GSSG ratio of liver. Simultaneously, histologic examination was also carried out with Hematoxylin-eosin staining (HE staining), Van Gieson staining (V.G staining) and fatty staining.

Results

Compared with the negative control group, CCl_4 dramatically increased serum ALT, AST, LDH, T-BIL, ALP, HBDH, ALB and CREA; decreased serum TC, HDL, LDL, BUN, and raised the A/G ratio; apparently elevated liver HYP and MDA, and reduced the GSH/GSSG ratio. Compared with the CCl_4 model group, orycho significantly decreased serum ALT, AST and LDH, reduced serum T-BIL, ALP, HBDH, ALB, CREA, obviously induced serum TC, elevated serum HDL and LDL slightly; significantly reduced liver HYP and decreased liver MDA to some degree. In addition, histologic changes also supported the biochemical results that orycho has obvious protective effect against the liver fibrosis and steatosis caused by CCl_4 in mice.

Conclusion

Orycho has protective effect on liver fibrosis and steatosis caused by CCl_4 in mice.

材料和方法

受试物

二月兰籽来源于北京。取二月兰籽 180g，于 2000mL 蒸馏水中煎煮，沸腾后 30min 后用 16 层纱布过滤；重复 2 次；取 3 次滤液共 6000mL 经加热浓缩至 100mL，制得二月兰籽水煎剂，每 1mL 水煎剂相当于二月兰籽生材量 1.8g。

动物分组及处理

120 只雄性 ICR 小鼠随机分为 6 组，每组 20 只，组 1 为溶剂对照组，组 2 为 CCl_4 模型组 [2%CCl_4- 橄榄油溶液（*V/V*），给药体积 5mL/kg，相当于 CCl_4 剂量 0.1mL/kg]，组 3 ~ 组 5 为二月兰籽水煎剂组（二月兰籽水煎剂 1.8kg/L，给药体积分别为 5mL/kg，10mL/kg 和 20mL/kg，相当于生药剂量 9g/kg，18g/kg，36g/kg），组 6 为二月兰籽水煎剂单独组（二月兰籽水煎剂 1.8kg/L，给药体积 20mL/kg，相当于二月兰籽剂量 36g/kg）。

实验周期为 16 周。CCl_4 给药前 1 周开始组 3~ 组 6 每日经口 ig 预防性给药相应剂量的二月兰籽水煎剂；第 1 周开始组 2~ 组 5 皮下注射 CCl_4- 橄榄油溶液每周 2 次（连续 2 天，周三和周四），同时组 3~ 组 6 每天经口 ig 给予相应计量的二月兰籽水煎剂；正常对照组经口 ig 给予等体积蒸馏水，皮下注射等体积橄榄油；连续给药 16 周。

血液生化指标测定

采用 HITACHI-7020 型自动生化分析仪测定血清中 ALT、TG、TC、LDL、HDL 的水平。

肝组织 MDA 含量测定（同酒精肝试验）

肝脏中 MDA 含量用 TBA 反应法检测（同酒精肝试验）。

留取小鼠肝脏组织后，用匀浆机制备肝脏组织匀浆，用羟脯氨酸试剂盒（碱 水解法）测定肝脏中 HYP 的含量。

数据处理

用 SPSS 18.0 统计学软件对本研究所得结果进行统计分析，数据用均数 ± 标准差表示，用单因素方差分析进行处理。

Materials and methods

Materials

Orycho seeds come from Beijing. 180g of orycho seeds were decocted in 2000mL distilled water, and filtered with 16 layers of gauze after boiling for 30min; Repeat twice; A total of 6000mL of filtrate from three times was heated and concentrated to 100mL, and the orycho seeds decoction was prepared. Every 1mL of water decoction was equivalent to 1.8g of raw orycho seeds.

Animal grouping and treatment

120 male ICR mice were randomly divided into 6 groups, with 20 mice in each group. Group 1 was the solvent control group, Group 2 was the CCl_4 model group [2% CCl_4- olive oil solution (*V/V*), the administration volume was 5mL/kg, which was equivalent to CCl_4 dose of 0.1mL/kg], and Group 3~ Group 5 was the orycho seeds decoction groups (18kg/L, the adminstration Volume were 5mL/kg, 10mL/kg and 20mL/kg, which were equivalent to the crude seed doses of 9g/kg, 18g/kg and 36g/kg). Group 6 was a single group of orycho seeds decoction (1.8kg/L, the administration volume was 20mL/kg, which was equivalent to the crude seeds of 36g/kg).

The experimental period is 16 weeks. One week before CCl_4 administration, group 3 ~ group 6 were given daily oral ig preventive administration of the corresponding dose of orycho seeds decoction; From the first week, CCl_4-olive oil solution was injected subcutaneously twice a week (for two consecutive days, Wednesday-Thursday) in groups 2~5, and at the same time, groups 3~6 were given a corresponding amount of orycho seeds decoction by oral ig every day; The normal control group was given equal volume distilled water by oral ig and injected with equal volume olive oil subcutaneously. It was administered continuously for 16 weeks.

Determination of blood biochemical indexes

The levels of alanine aminotransferase (ALT), triglyceride (TG), total cholesterol (TC), low density lipoprotein (LDL) and high density lipoprotein (HDL) in serum were measured by HITACHI-7020 automatic biochemical analyzer.

Determination of MDA content in liver tissue (same as alcoholic hepatitis test)

MDA content in liver was detected by TBA reaction method (same as alcoholic hepatitis test).

After the mouse liver tissue was collected, the liver homogenate was prepared by a homogenizer, and the content of hydroxyproline (HYP) in the liver was determined by hydroxyproline kit (alkaline hydrolysis).

Data processing

SPSS 18.0 was used to analyze the results of this study. The data were expressed by mean ± standard deviation and processed by one-way analysis of variance.

结果

Results

二月兰籽对 CCl_4 所致肝纤维化小鼠脏器重及脏器系数的影响

与正常对照组相比，小鼠在皮下注射 CCl_4 16 周后，体重变化不明显，肝重及肝体比升高近10%。二月兰籽对肝体比变化不明显（表 4-6）。

Effects of orycho Seed on Organ Weight and Organ Coefficient of CCl_4-induced Hepatic Fibrosis Mice

Compared with the normal control group, after 16 weeks of subcutaneous injection of CCl_4, the body weight of the mice did not change significantly, and the liver weight and liver-to-body ratio were increased by nearly 10%. The liver- to-body ratio of orycho seeds was not significantly changed (Table 4-6).

表 4-6 小鼠体重和肝重变化

Table 4-6 Mouse body weight and liver weight changes

分组 Groups （g/kg）	体重 Body (g)	肝 Liver (g)	肝 / 体 L/B (%)
正常对照 Normal control	49.54±3.87	2.42±0.45	4.87±0.61
CCl_4	49.31±2.29	2.68±0.24##	5.44±0.39##
二月兰籽 -9+CCl_4 Orycho seeds-9+CCl_4	43.89±2.63##	2.28±0.26**	5.20±0.45*
二月兰籽 -18+CCl_4 Orycho seeds-18+CCl_4	44.55±3.42##	2.53±0.31	5.68±0.43
二月兰籽 -36+CCl_4 Orycho seeds-36+CCl_4	40.99±2.36**	2.40±0.25**	5.85±0.42**

注：数据为均数 ± 标准差，n=20。##$P<0.01$（与对照组比较）；*$P<0.05$，**$P<0.01$（与 CCl_4 组比较）。

Data is expressed by mean ± SEM. n=20 . ##$P<0.01$, compare with Normal Group; *$P<0.05$, **$P<0.01$, compare with CCl_4 Group.

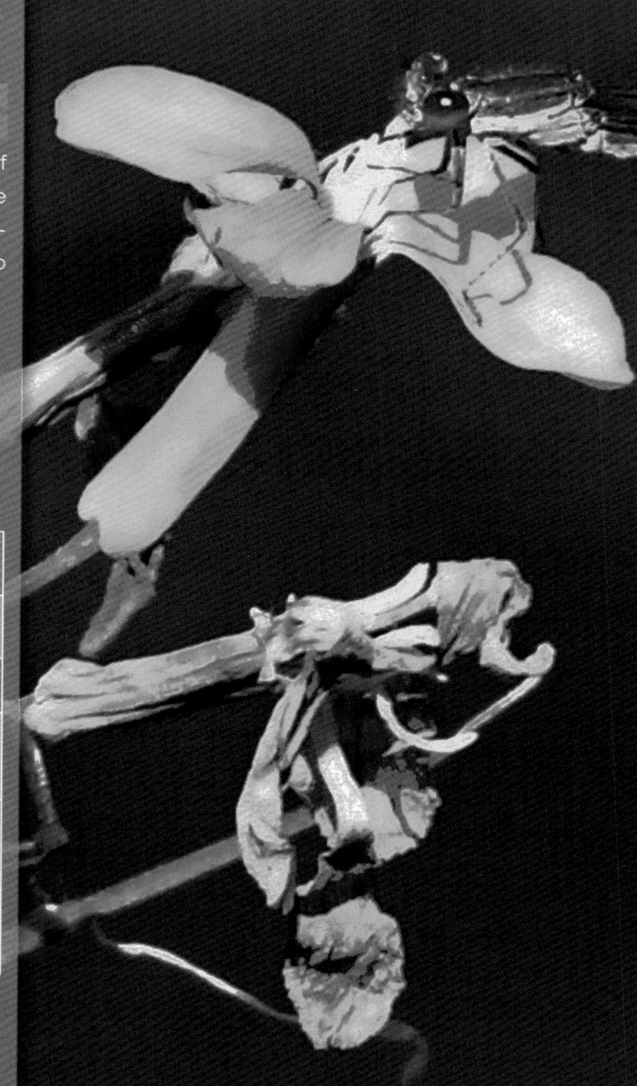

二月兰籽及 CCl_4 对小鼠血清 ALT、AST 及 LDH 含量的影响

与正常对照组相比，小鼠皮下注射 CCl_4 16 周后，血清中的生化指标水平都明显升高。与 CCl_4 模型组相比，二月兰籽各剂量组小鼠血清指标随着给药剂量的增加，下降程度也随之加大。与正常对照组相比，二月兰籽水煎剂单独组小鼠血清中 T-BIL 含量升高约 14%（$P<0.05$），其他指标没有明显差异（表 4-7）。

Effects of Orycho Seed and CCl_4 on Serum ALT, AST and LDH Contents in Mice

After 16 weeks of subcutaneous injection of CCl_4, compared with the normal control group, the levels of the biochemical indicators in the serum of the mice were significantly increased. Compared with the CCl_4 model group, the serum indicators of mice in each dosage group of orycho seed decreased more severely with the increase of the administration dose. Compared with the normal control group, the serum T-Bil content of mice in the orycho alone group was increased by about 14% (P<0.05), and there was no significant difference in other indicators (Table 4-7).

表 4-7 二月兰抗长期 CCl_4 所致小鼠肝功酶的变化

Table 4-7 Effect of orycho on CCl_4 induced changes of liver functions

分组 Groups (g/kg)	ALT (U/L)	AST (U/L)	LDH (U/L)	ALP (U/L)	HBDH (U/L)	T-BIL (U/L)
正常对照 Normal control	44±8	99±24	710±153	44.08±9.86	216.82±79.28	1.43±0.25
CCl_4	1070±153##	476±152##	1588±452##	58.15±12.38##	373.45±110.4##	2.26±1.12#
二月兰籽 -9+CCl_4 Orycho seeds-9+CCl_4	694±349	289±118	1147±440	56.71±11.43	332.9±98.7	2.18±0.94
二月兰籽 -18+CCl_4 Orycho seeds-18+CCl_4	655±361*	318±88	1297±207	52.53±14.94	365.78±74.97	2.12±1.08
二月兰籽 -36+CCl_4 Orycho seeds-36+CCl_4	503±358*	261±95*	1012±115*	49.25±14.88*	330.77±89.82	2.13±0.86
二月兰籽 -36 Orycho seeds-36	54±11	121±24	571±106	49.69±7.89	189.8±42.43	1.84±0.47#

注：数据为均数 ± 标准差，n=20。#$P<0.05$，##$P<0.01$，（与对照组比较）；*$P<0.05$（与模型组比较）。

Data is expressed by mean ± SEM. n=20. #$P<0.05$, ##$P<0.01$, compare with Normal Group; *$P<0.05$, compare with CCl_4 Group.

二月兰籽及 CCl_4 对小鼠血清 TP、ALB 及白球比（A/G）的影响

与正常对照组相比，皮下注射 $CCl_4$16 周后，小鼠血清 ALB 水平升高较明显，升高约 10%，A/G 比值也升高约 15%。但是，CCl_4 尚未对肝脏的蛋白合成能力产生明显的影响。与 CCl_4 模型组相比，二月兰籽水煎剂各剂量组小鼠血清中 TP 和 ALB 水平变化均不明显，A/G 比值也无明显差异。与正常对照组相比，A/G 比值无明显差异（表 4-8）。

Effects of orycho Seed and CCl_4 on Serum TP, Alb and White-ball Ratio (A/G) in Mice

After 16 weeks of subcutaneous injection of CCl_4, compared with the normal control group, the serum ALB level of the mice increased significantly by about 10%, and the A/G ratio also increased by about 15%. Compared with the CCl_4 model group, the serum levels of TP and ALB and the A/G ratio were not significantly changed in all the dosage groups. The A/G ratio was not significantly different from the normal control group (Table 4-8).

表 4-8 小鼠血清 TP、ALB 及 A/G 的变化
Table 4-8 Changes of serum TP、Alb levels and A/G ratio in mice

分组 Groups（g/kg）	TP (U/L)	ALB (U/L)	A/G
正常对照 Normal control	55.79±3.49	27.99±2.75	1.01±0.12
CCl_4	57.69±5.58	30.57±2.52##	1.15±0.17##
二月兰籽 -9+CCl_4 Orycho seeds-9+CCl_4	57.69±1.79	31.19±0.99	1.18±0.12
二月兰籽 -18+CCl_4 Orycho seeds-18+CCl_4	56.22±2.55	30.38±2.40	1.18±0.15
二月兰籽 -36+CCl_4 Orycho seeds-36+CCl_4	58.30±3.43	31.74±1.45	1.20±0.10
二月兰籽 -36 Orycho seeds-36	60.32±1.50##	31.26±1.26##	1.04±0.07

注：数据为均数 ± 标准差，n=20。##P<0.01,（与对照组比较）。
Data is expressed by mean ± SEM. n=20. ##P<0.01, compare with Normal Group.

二月兰籽及 CCl_4 对小鼠血清 TC、TG、HDL 及 LDL 含量的影响

与正常对照组相比，给予 CCl_4 16 周后，小鼠血清 TC 和 HDL 水平分别下降 31% 和 44%，LDL 水平下降约 50%，TG 水平变化不大。小鼠血清中 TC 含量明显下降，TG 含量稍有下降，说明肝细胞损伤严重，肝脏脂类代谢发生障碍。与 CCl_4 模型组相比，二月兰籽水煎剂各剂量组小鼠血清中 TC 和 LDL 水平有所升高，且随着给药剂量的增加，升高程度也随之增加。二月兰籽水煎剂各剂量 组小鼠血清中 TG 水平下降明显，HDL 水平稍有升高与正常对照组 相比，二月兰水煎剂单独组小鼠血清中 TC 水平升高，TG 水平下 降，HDL 和 LDL 水平变化不大（表 4-9）。

Effects of Orycho Seed and CCl_4 on Serum TC, TG, HDL and LDL Contents in Mice

After 16 weeks of administration of CCl_4, compared to the normal control group, mouse serum TC and HDL levels decreased by 31% and 44%, respectively, LDL levels decreased by approximately 50%, and TG levels remained unchanged. Compared with the CCl_4 model group, the serum TC and LDL levels of mice in each dose group of orycho seeds decoction were increased, and the degree of increase was also increased with the increase of the dose. Serum TG level was decreased significantly and HDL level was slightly increased in each dosage group of orycho seeds decoction. Compared with the normal control group, serum TC level was increased, TG level was decreased, and HDL and LDL levels were not changed significantly in the mice of orycho seeds decoction group (Table 4-9).

表 4-9 小鼠血清 TC、TG、HDL 和 LDL 水平的变化

Table 4-9 Changes of serum TC, TG, HDL and LDL levels in mice

分组 Groups （g/kg）	TC （U/L）	TG （U/L）	HDL （U/L）	LDL （U/L）
正常对照 Normal control	4.51±0.67	1.79±0.52	2.12±0.59	0.20±0.26
CCl_4	3.12±0.60$^{##}$	1.73±0.52	1.18±0.39$^{##}$	0.10±0.04$^{#}$
二月兰籽 -9+CCl_4 Orycho seeds-9+CCl_4	3.24±0.50	1.26±0.42**	1.26±0.43	0.10±0.04
二月兰籽 -18+CCl_4 Orycho seeds-18+CCl_4	3.23±0.47	1.22±0.40**	1.18±0.39	0.11±0.04
二月兰籽 -36+CCl_4 Orycho seeds-36+CCl_4	3.82±0.89**	1.19±0.29**	1.29±0.56	0.16±0.11**
二月兰籽 -36 Orycho seeds-36	5.14±0.34$^{##}$	1.15±0.31**	2.34±0.25	0.24±0.08

注：数据为均数 ± 标准差，n=20。$^{#}P$<0.05，$^{##}P$<0.01,（与对照组比较）；$^{**}P$<0.01（与模型组比较）。
Data is expressed by mean ± SEM. n=20. $^{#}P$<0.05, $^{##}P$<0.01, compare with Normal Group; $^{**}P$<0.01, compare with CCl_4 Group.

小鼠肝组织 MDA 含量及 GSH/GSSG 比值的影响

与正常对照组相比，给予 CCl_4 16 周后，小鼠肝组织 MDA 含量升高近 1 倍，GSH/GSSG 比值降低约 20%。与 CCl_4 模型组相比，二月兰籽水煎剂各剂量组小鼠肝组织 MDA 含量分别下降 23%、15% 和 20%，肝组织 GSH/GSSG 比值升高不明显。与正常对照组相比，二月兰籽水煎剂单独组小鼠肝组织中 MDA 含量未见明显差异；GSH/GSSG 变化不明显（表 4-10）。

Effects of MDA content and GSH/GSSG ratio in mouse liver tissue

After 16 weeks of administration of CCl_4, compared with the normal control group, the MDA content in the liver tissue of the mice was nearly doubled and the GSH/GSSG ratio was decreased by about 20%. Compared with the CCl_4 model group, the MDA content in the liver tissue of each dose group of orycho seeds decoction was decreased by 23%, 15%, and 20%, respectively, and the GSH/GSSG ratio in the liver tissue was not significantly increased. Compared with the normal control group, there was no significant difference in the MDA content in the liver tissue of mice in the orycho seeds decoction group (Table 4-10).

表 4-10 二月兰抗长期 CCl_4 所致小鼠肝鼠 MDA、GSH/GSSG 比值的变化

Table 4-10 Effect of orycho on CCl_4 induced changes of liver MDA、GSH/GSSG

分组 Groups （g/kg）	MDA （μmol/g liver)	GSH/GSSG
正常对照 Normal control	0.34±0.20	5.16±1.33
CCl_4	0.66±0.47#	4.19±1.02##
二月兰籽 -9+CCl_4 Orycho seeds-9+CCl_4	0.51±0.28	4.21±1.06
二月兰籽 -18+CCl_4 Orycho seeds-18+CCl_4	0.56±0.29	4.27±0.46
二月兰籽 -36+CCl_4 Orycho seeds-36+CCl_4	0.53±0.33	4.17±0.34
二月兰籽 -36 Orycho seeds-36	0.35±0.24	4.52±0.55

注：数据为均数 ± 标准差，n=20。$^{\#}P$ < 0.05，$^{\#\#}P$ < 0.01,（与对照组比较）。

Data is expressed by mean ± SEM. n=20. $^{\#}P$<0.05, $^{\#\#}P$< 0.01, compare with Normal Group.

二月兰籽抗肝硬化的作用

与正常对照组相比，皮下注射 CCl_4 16 周后，小鼠肝组织 HYP 含量明显升高，升高幅度达 91%。与 CCl_4 模型组相比，二月兰籽水煎剂各剂量组小鼠肝组织中 HYP 含量明显下降，且随着给药剂量的增加，下降程度也随之增加。与正常对照组相比，二月兰籽水煎剂单独组小鼠肝组织 HYP 含量没有明显差异（表 4-11）。

Effect of orycho seed on anti-liver cirrhosis

After 16 weeks of subcutaneous injection of CCl_4, compared with the normal control group, the HYP content in the liver tissue of the mice was significantly increased, by 91%. Compared with the CCl_4 model group, the HYP content in the liver tissue of mice in each dose group of orycho seeds decoction was decreased significantly, and the degree of decline was increased with the increase of the dose. Compared with the normal control group, there was no significant difference in the HYP content in the liver tissue of the mice in the group treated with orycho seeds decoction alone (Table 4-11).

表 4-11 小鼠肝 HYP 的变化

Table 4-11 Changes of HYP in Liver of Mice

分组 Groups（g/kg）	HYP (μg/mg liver)
正常对照 Normal control	0.11±0.02
CCl_4	0.21±0.04##
二月兰籽 -9+CCl_4 Orycho seeds-9+CCl_4	0.21±0.03
二月兰籽 -18+CCl_4 Orycho seeds-18+CCl_4	0.18±0.04**
二月兰籽 -36+CCl_4 Orycho seeds-36+CCl_4	0.13±0.01**
二月兰籽 -36 Orycho seeds-36	0.10±0.03

注：数据为均数 ± 标准差，n=20。##P<0.01,（与对照组比较）；**P < 0.01（与模型组比较）。

Data is expressed by mean ± SEM. n=20. ##P<0.01, compare with Normal Group; **P<0.01, compare with CCl_4 Group.

肝组织病理变化（图 4-3）

取染色后切片的相同位置进行光学显微镜观察发现，正常对照组 HE 染色（图 4-3 A1）显示肝小叶结构清晰，胞膜完整，细胞核大而圆，胞质丰富，肝细胞无变性坏死，肝细胞索排列整齐；VG 染色（图 4-3 A2）显示汇管区和小叶间无胶原纤维形成。

CCl_4 模型组 HE 染色（图 4-3 B1）呈现明显的弥漫性肝细胞坏死，肝细胞正常结构被破坏，肝细胞界限不清楚，胞质疏松呈网状，细胞核固缩或裂解，肝细胞索排列紊乱，肝组织损伤严重；VG 染色（图 4-3 B2）显示汇管区胶原纤维形成明显，并向小叶延伸，小叶结构紊乱。

二月兰籽各剂量组 HE 染色（图 4-3 C1、图 4-3 D1、图 4-3 E1）显示肝组织细胞及结构破坏程度明显减轻，肝细胞变性、坏死减少，且随着二月兰籽给药剂量的增加，肝组织病变进一步减轻，高剂量组肝细胞结构明显；VG 染色（图 4-3 C2、图 4-3 D2、图 4-3 E2）显示肝细胞损伤明显减轻，胶原纤维较少或不明显。

二月兰籽单独组 HE 染色（图 4-3 F1）未见肝组织结构破坏；VG 染色（图 4-3 F2）与阴性对照组无明显差别。

Pathological changes of liver tissue (Fig.4-3)

HE staining (Fig. 4-3 A1) in the normal control group showed that the hepatic lobular structure was clear, and the cell membranes were intact. The nuclei were large and round, with abundant cytoplasm. There was no degeneration and necrosis of hepatocytes, and the hepatocyte cords were arranged neatly. VG staining (Fig.4-3 A2) showed no collagen fiber formation in the portal area and between the lobules.

HE staining (Fig.4-3 B1) in the CCl_4 model group showed obvious diffuse hepatocyte necrosis, the normal structure of hepatocytes was damaged, the boundary of hepatocytes was unclear, the cytoplasm was loose and reticular, the nuclei were pyknotic or lysed, the hepatocyte cords were arranged disorderly, and the liver tissue was seriously damaged. VG staining (Fig.4-3 B2) showed clear collagen fiber formation in the portal area and extending to the lobules, with disorganized lobular structure.

HE staining (Fig.4-3 C1, Fig. 4-3 D1, Fig.4-3 E1) showed that the degree of cell and structure damage of liver tissue was significantly reduced, and the degeneration and necrosis of liver cells were reduced in each dose group of Cymbidium goeringii. With the increase of the dose of orycho seed, the lesions of liver tissue were further alleviated, and the structure of liver cells in the high dose group was significant. VG staining (Fig.4-3 C2, Fig.4-3 D2, Fig.4-3 E2) showed a significant reduction in hepatocyte injury with few or no clear collagen fibers.

HE staining (Fig.4-3 F1) showed no structural damage to the liver in the cymbidium goeringii seed alone group. VG staining (Fig.4-3 F2) was not significantly different from that in the negative control group.

图 4-3 各组小鼠肝脏组织病理变化（×200）

Fig.4-3 Histopathologic changes caused by CCl_4 in the liver of mice by HE and VG staining (×200)

A—正常对照；B—CCl_4；C—二月兰籽-9+CCl_4；D—二月兰籽-18+CCl_4；E—二月兰籽-36+CCl_4；F—二月兰籽-36:F1: HE 染色；F2: VG 染色。黑箭头为胶原纤维

A—Normal control; B—CCl_4; C—orycho seeds-9+CCl_4; D—orycho seeds-18+CCl_4; E—orycho seeds-36+CCl_4; F—orycho seeds-36: F1, HE staining; F2, VG staining. The black arrow shows the collagenous fiber

图 4-4 各组小鼠肝脏组织病理变化（苏丹Ⅲ染色；×200）
A—正常对照；B—CCl_4；C—二月兰籽 -9+CCl_4；D—二月兰籽 -18+CCl_4；E—二月兰籽 -36+CCl_4；F—二月兰籽 -36

Fig.4-4 Histopathologic fate changes caused by CCl_4 in the liver of mice by Sudan Ⅲ staining (×200)
A—Normal control; B—CCl_4; C—orycho seeds-9+CCl_4; D—orycho seeds-18+CCl_4; E—orycho seeds-36+CCl_4; F—orycho seeds-36

肝脏脂肪变化（图 4-4）

对肝脏进行冰冻切片后进行脂肪染色，脂肪染色使脂滴染橘黄色，细胞核染蓝色。正常对照组（图 4-4 A）肝小叶结构清晰，细胞核大而圆，汇管区周围和小叶间均无脂肪变性发生。CCl_4 模型组（图 4-4 B）肝细胞索排列紊乱，汇管区周围和小叶间有大量脂滴分布，脂肪变性范围大、发生明显。二月兰籽各剂量组（图 4-4 C、图 4-4 D、图 4-4 E）肝细胞脂肪变性明显减轻，只周围散在少量脂滴。二月兰籽单独组（图 4-4 F）与阴性对照组无明显差别。

Liver fat changes (Fig. 4-4)

Frozen sections of the liver were stained with fat, which stained the fat droplets orange and the nuclei blue. In the normal control group (Fig.4-4 A), the hepatic lobular structure was clear, and the nuclei were large and round. There was no occurrence of fatty degeneration around the portal area and between the lobules. In the CCl_4 model group (Fig.4-4 B), the hepatocyte cords were arranged disorderly, a large number of lipid droplets were distributed around the portal area and between the lobules, and the steatosis was large in scope and obvious. The fatty degeneration of hepatocytes was significantly reduced and only a small number of lipid droplets were scattered around in the (Fig.4-4 C, Fig.4-4 D, Fig.4-4 E) hepatocytes of each dose group of Oroxylum seeds. There was no significant difference between the Cymbidium goeringii seed group (Fig.4-4 F) and the negative control group.

计算机电子照相系统拍照后用 Photoshop 7.0 图像处理软件对脂肪染色病理图片进行脂肪变、坏死和变性组织面积半定量分析。其中，正常对照组未见明显病理损伤；CCl_4 模型组肝组织损伤面积最大，占总面积的 63.64% ± 14.61%（$P < 0.01$）；二月兰籽水煎剂各组肝组织损伤面积均较 CCl_4 模型组减少，且随着剂量的增加，肝组织损伤面积减少的程度更明显。各剂量组肝组织损伤面积分别为 40.24% ± 6.91%（$P < 0.05$）、18.28% ± 6.31%（$P < 0.01$）和 4.73% ± 5.58 %（$P < 0.01$）。水煎剂单独租未见明显病理损伤（表 4-12）。

After taking pictures by computer electronic photography system, the tissue area of fatty change, necrosis and degeneration was analyzed by Photoshop 7.0 image processing software. Among them, no obvious pathological damage was found in the normal control group. The injured area of liver tissue in CCl_4 model group was the largest, accounting for 63.64%±14.61% of the total area (P<0.01). Compared with CCl_4 model group, the damaged area of liver tissue in each group of orycho seeds decoction decreased significantly with the increase of dosage. The damaged area of liver tissue in each dose group was 40.24%±6.91% (P<0.05),18.28%±6.31% (P<0.01) and 4.73% ±5.58% (P<0.01), respectively. No obvious pathological damage was found in orycho seeds decoction control group (Table 4-12).

表 4-12 小鼠肝脏脂肪变等病理变化面积

Table 4-12 Changes of histopathologic damage area in mouse liver

分组 Groups（g/kg）	损伤面积 Damaged area（%）
正常对照 Normal control	0.00 ± 0.00
CCl_4	63.64 ± 14.61##
二月兰籽 -9+CCl_4 Orycho seeds-9+CCl_4	40.24 ± 6.91*
二月兰籽 -18+CCl_4 Orycho seeds-18+CCl_4	18.28 ± 6.31**
二月兰籽 -36+CCl_4 Orycho seeds-36+CCl_4	4.73 ± 5.58**
二月兰籽 -36 Orycho seeds-36	0.00 ± 0.00

注: 数据为均数 ± 标准差。## P<0.01,（与对照组比较）; * P<0.05，** P<0.01（与模型组比较）。

Data is expressed by mean ± SEM. # P<0.05, ## P<0.01, compare with normal control; group; * P <0.05, ** P<0.01, compare with CCl_4 group.

讨论

肝纤维化以胞外基质大量增生并沉积于肝脏为主要特征，以胶原为 主要成分。各种致病因子可直接或间接地通过细胞因子刺激细胞基因转录，从而使胶原mRNA增加， 导致胶原合成增加。羟脯氨酸（HYP）是一种非必需氨基酸，是胶原代谢的重要产物。HYP含量的测定已成为衡量机体胶原组织代谢的重要指标，能客观地反映肝纤维化程度，是评价肝纤维化程度的生物学标志。

实验结果显示，CCl_4 染毒 16 周后，小鼠肝组织 HYP 含量明显升高，说明肝脏胶原含量升高，小鼠出现肝纤维化。二月兰水煎剂可阻止小鼠 CCl_4 肝组织引起的 HYP 上升，且随着给药剂量的升高，作用越显著，说明能抑制 CCl_4 所致小鼠肝纤维化。

与正常对照组相比，皮下注射 CCl_4 16 周后，小鼠肝组织 MDA 含量明显升高，GSH/GSSG 比值明显下降，说明 CCl_4 引起了脂质过氧化损伤。而与模型组相比，二月兰籽能一定程度上降低肝组织 MDA 含量，并恢复 GSH/GSSG 比值，说明二月兰籽对 CCl_4 引起的脂质过氧化损伤产生了一定的保护作用。

CCl_4 能够造成肝细胞索排列紊乱，汇管区周围和小叶间有大量脂滴分布，脂肪变性范围大而且 MDA 含量明显升高。说明长期给药 CCl_4 会导致肝细胞发生脂肪变等变性和坏死，细胞间胶原纤维增多，最终造成肝纤维化。二月兰籽能明显抑制肝细胞坏死及胶原纤维的形成，对于肝纤维化和脂肪变具有一定的保护作用。

Discussion

Hepatic fibrosis is mainly characterized by the massive proliferation of extracellular matrix and its deposition in the liver, with collagen as the main component. Various pathogenic factors can stimulate transcription of cellular genes directly or indirectly through cytokines, thereby increasing collagen mRNA and leading to increased collagen synthesis. Hydroxyproline (HYP) is a non-essential amino acid and is an important product of collagen metabolism. The measurement of HYP content has become an important indicator of collagen tissue metabolism, which can objectively reflect the degree of liver fibrosis. Is a biological marker for evaluating the degree of liver fibrosis.

The experimental results showed that the HYP content in the mouse liver tissue was significantly increased after 16 weeks of exposure to CCl_4, indicating that the collagen content in the liver was increased, and the mouse developed liver fibrosis. The orycho seeds decoction could prevent the HYP increase caused by CCl_4 in mouse liver tissue, and the more obvious the effect was with the increase of the dose, indicating that it could inhibit CCl_4-induced liver fibrosis in mice.

Compared with the normal control group, after 16 weeks of subcutaneous injection of CCl_4, the MDA content in the liver tissue of the mice was significantly increased, and the GSH/GSSG ratio was significantly decreased, indicating that CCl_4 caused lipid peroxidation damage. Compared with the model group, the seeds of orycho could reduce the MDA content in the liver tissue to a certain extent and restore the GSH/GSSG ratio, indicating that the seeds of orycho had a certain protective effect on CCl_4-induced lipid peroxidation.

CCl_4 can cause disorganized arrangement of hepatocyte cords, distribution of a large number of lipid droplets around portal areas and between lobules, a large range of steatosis and significantly increased MDA content. These results indicated that long-term administration of CCl_4 would lead to steatosis, isochronous degeneration and necrosis of hepatocytes, and increase in intercellular collagen fibers, finally leading to liver fibrosis. orycho seed could significantly inhibit hepatocyte necrosis and collagen fiber formation, and had a certain protective effect on liver fibrosis and fat liver.

结论

根据上述实验结果可以发现，二月兰籽水煎剂能明显缓解 CCl_4 引起的肝硬化和脂肪肝等慢性肝损伤，且随着给药剂量的增加，保护作用逐渐增强。

Conclusion

According to the above experimental results, it could be found that orycho seeds decoction could significantly alleviate chronic liver injury such as liver cirrhosis and fatty liver caused by CCl_4, and the protective effect was gradually enhanced with the increase of the dose.

二月兰的新保肝成分

背景

我们发现二月兰具有保肝的解毒剂作用后，找到二月兰的有效成分就成了我们的一项重要任务。

New liver-protecting components of orycho

Background

After we found that orycho has the antidotes of protecting liver, it became an important task for us to find the effective components of orycho.

摘要

目的

对二月兰籽水煎剂中抗中毒性肝炎的成分进行分离和提纯，检测其中的有效成分。

方法

对二月兰籽的有效成分进行分离提纯，采用大孔树脂乙醇梯度洗脱、反相高效液相色谱法、凝胶分离方法和半制备高效液相色谱法对二月兰籽水煎剂的物质成分进行分离提取。采用雄性ICR小鼠，以二月兰籽水煎剂，及其分离物为受试药物，采用CCl_4肝损伤模型，测量小鼠血清中肝功酶丙氨酸氨基转移酶、天冬氨酸氨基转移酶和乳酸脱氢酶的含量，观察肝病理变化，确定保肝的有效成分。采用超高效液相色谱法和核磁共振法以及单晶衍射法等对有效单体化合物进行结构测定和含量测定。

结果

二月兰籽水煎剂及其分离提取出的抗中毒性肝炎有效成分新化合物能显著降低CCl_4肝炎症小鼠肝功酶的活性升高和组织学病理的改变。有效成分是一种新的天然化合物，我们取名为二月兰苷；在二月兰籽水煎剂干粉中的含量为0.0529%。

结论

我们找到了二月兰籽中的主要抗中毒性肝炎的有效成分是新化合物二月兰苷。

Abstract

Purpose

To separate and purify the anti-toxic hepatitis components in the orycho seeds decoction, and to detect the effective components.

Method

The effective components of the seeds of orycho were separated and purified, and the components of the orycho seeds decoction were separated and extracted by macroporous resin ethanol gradient elution, reversed-phase high performance liquid chromatography, gel separation and semi-preparative high performance liquid chromatography. Male ICR mice were used, and the orycho seeds decoction and its separation were used as the test drugs. The liver injury model of CCl_4 was adopted. The contents of liver enzymes alanine aminotransferase, aspartate aminotransferase and lactate dehydrogenase in mice serum were measured, and the pathological changes of liver diseases were observed to determine the effective components of liver protection. The structure and content of effective monomer compounds were determined by ultra-high performance liquid chromatography, nuclear magnetic resonance and single crystal diffraction.

Results

The new compound of anti-toxic hepatitis extracted from the orycho seeds decoction can significantly reduce the increase of liver function enzyme activity and histological and pathological changes in mice with liver inflammation induced by CCl_4. The active ingredient is a new natural compound, which is named as orychonin. Its content in the dry powder of the orycho seeds decoction is 0.0529%.

Conclusion

We found that the main effective component of anti-toxic hepatitis in the seeds of orycho is the new compound orychonin.

技术路线

见图 4-5。

Technical pipeline

see Fig.4-5.

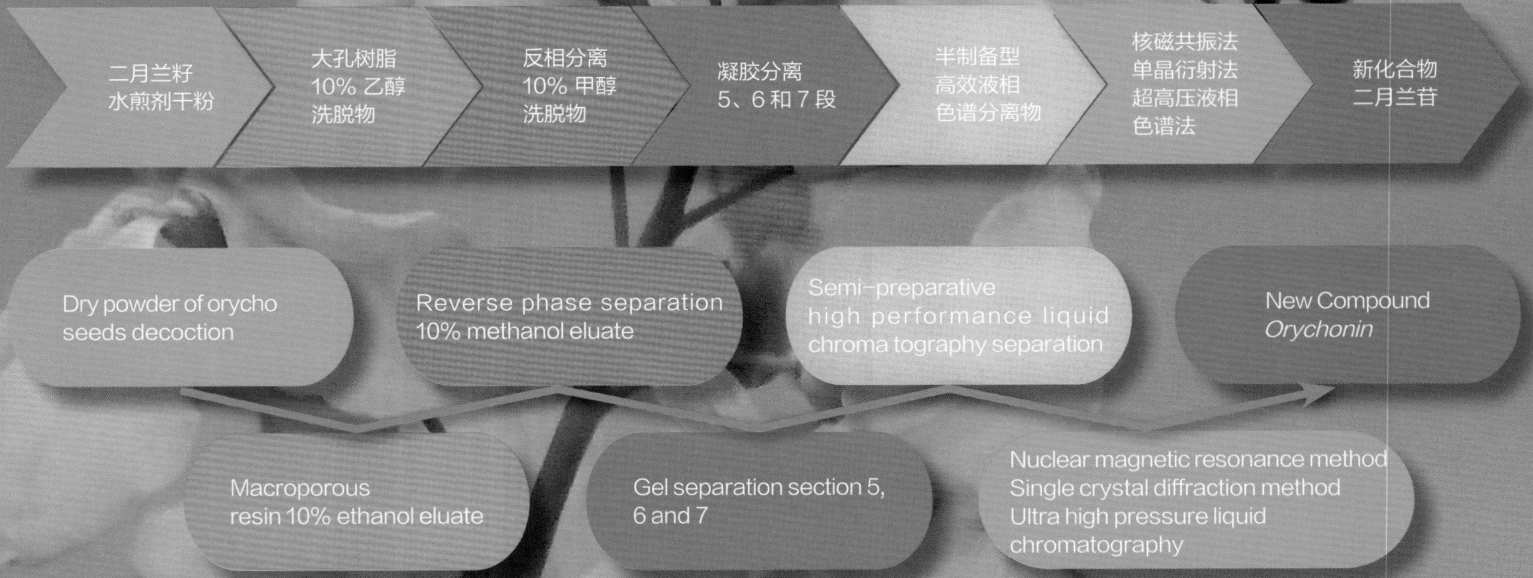

二月兰籽水煎剂干粉 → 大孔树脂 10% 乙醇洗脱物 → 反相分离 10% 甲醇洗脱物 → 凝胶分离 5、6 和 7 段 → 半制备型高效液相色谱分离物 → 核磁共振法 单晶衍射法 超高压液相色谱法 → 新化合物 二月兰苷

Dry powder of orycho seeds decoction → Macroporous resin 10% ethanol eluate → Reverse phase separation 10% methanol eluate → Gel separation section 5, 6 and 7 → Semi-preparative high performance liquid chroma tography separation → Nuclear magnetic resonance method Single crystal diffraction method Ultra high pressure liquid chromatography → New Compound *Orychonin*

图 4-5 五大工程寻找二月兰有效化合物
Fig.4-5 Five major procedures to find effective compounds of orycho

二月兰籽水煎剂干粉的保肝作用

方法

二月兰籽水煎剂干粉制备：二月兰籽 50kg，10 倍蒸馏水煎煮 2 次，每次 1h，滤液合并，鼓风干燥箱干燥至膏状，真空干燥箱干燥至干粉，蒸馏水配成 9g/kg、18g/kg、36g/kg、72g/kg、144g/kg 的水煎剂干粉剂量组，灌胃剂量 0.2mL/10g。

造模：CCl_4（1.5% 溶于橄榄油），皮下注射 0.05mL/10g。

健康 ICR 小鼠（清洁级）40 只，雄性，体重为 20~22g，随机分为正常对照组和 CCl_4 模型组，二月兰籽水煎剂组，二月兰籽水煎剂干粉剂量组。连续给药 4 天，第 4 天皮下注射 CCl_4，24h 后取血测 ALT、AST 和 LDH，肝组织切片做 HE 染色。

Liver-protective activity of dry powder of orycho seed decoction

Methods

The preparation of dry powder of orycho seeds decoction was as follows: 50kg of orycho seeds were decocted with 10 times distilled water for 2 times, each time for 1h, the filtrates were combined, dried to paste in a blast drying oven, dried to dry powder in a vacuum drying oven, and distilled water was prepared into dry powder dosage groups of decoction of 9g/kg, 18g/kg, 36g/kg, 72g/kg and 144g/kg, with a dose of 0.2mL/10g for intragastric administration.

Modeling: CCl_4 (1.5% dissolved in olive oil) was injected subcutaneously at 0.05mL/10g.

Forty healthy ICR mice (clean grade),male,weighing 20~22g, were randomly divided into normal control group, CCl_4 model group, decoction group and dry powder dosage groups of orycho seeds decoction. After four days of continuous administration, CCl_4 was injected subcutaneously on the fourth day. After 24 hours, blood samples were taken to measure ALT, AST and LDH, and liver tissue sections were stained with HE method.

结果

（1）肝脏病理学结果

正常对照组（图 4-6 A）肝小叶结构清晰，肝细胞索围绕中央静脉成放射状排列，细胞结构完整清晰，无坏死、变性或炎症细胞浸润。CCl_4 模型组（图 4-6 B）的肝组织可见炎症细胞浸润，以淋巴细胞为主，中央静脉周围可见不同程度的充血，肝细胞可见不同程度的肿胀、坏死，细胞质水分含量增多，几乎透明，呈现出气球样变。二月兰籽水煎剂组（图 4-6 C）肝结构和正常对照组基本一致，未观察到明显病理改变。干粉对肝脏的改善有剂量效应关系。相同剂量的水煎剂和干粉的作用基本一致（图 4-6）。

（2）二月兰籽水煎剂干粉抗小鼠 CCl_4 肝损伤

在二月兰籽水煎剂、二月兰籽号干粉剂量组中，小鼠血清中的肝功酶 ALT、AST、LDH 水平趋近于正常对照组，与 CCl_4 模型组相比较，差异有统计学意义（$P<0.05$）。结果见表 4-13。

Results

(1) Liver pathology result

In the normal control group (Fig.4-6 A), the hepatic lobule structure was clear, the hepatic cords were arranged radially around the central vein, the cellular structure was complete and clear, and there was no necrosis, degeneration or inflammatory cell infiltration. Inflammatory cells infiltrated in the liver tissue of CCl_4 model group (Fig.4-6 B), mainly lymphocytes, with different degrees of congestion around the central vein, swelling and necrosis of liver cells, increased cytoplasm water content, almost transparent and balloon-like changes. The liver structure of the decoction group (Fig.4-6 C) was basically the same as that of the normal control group, and no obvious pathological changes were observed. There is a dose-effect relationship between dry powder and liver improvement. The effects of the same dose of decoction and dry powder are the same (Fig.4-6) .

(2) Anti-CCl_4 induced liver injury by dry powder of orycho seeds decoction in mice

The levels of liver function enzymes ALT, AST and LDH in the serum of mice in the dosage group of orycho seeds decoction and dry powder of orycho seeds were close to those in the normal control group, and the difference was statistically significant compared with the CCl_4 model group ($P<0.05$). The results are shown in table 4-13.

图 4-6 二月兰籽水煎剂干粉对小鼠 CCl_4 肝病理的影响

HE 染色，×200。A—正常组；B—CCl_4；C—水煎剂 -36； D—干粉 -9；E—干粉 -18；F—干粉 -36；G—干粉 -72； H—干粉 -144

Fig.4-6 Effect of dry powder of orycho seeds decoction on liver pathology of CCl_4 in mice

HE method, ×200. A—Control; B—CCl_4; C—Decoction-36; D—Dry powder-9; E—Dry powder-18; F—Dry powder-36; G—Dry powder-72; H—Dry powder-144

表 4-13 二月兰籽水煎剂干粉对血清肝功酶的影响

Table 4-13 Effect of dry powder of orycho seeds decoction on CCl_4 induced serum liver function enzyme changes

分组 Groups（g/kg）	死亡 Death	ALT（U/L）	AST（U/L）	LDH（U/L）
正常对照组 Normal control	0/5	39.80±7.19	131.20±61.77	951.00±330.67
CCl_4	0/5	854.50±377.84#	577.5±310.08#	1451.83±756.35
水煎剂 -36 Decoction-36	0/5	41.80±4.76*	147.40±29.10*	1009.60±162.85
干粉 -9 Dry powder-9	0/5	45.40±17.11*	171.4±26.35*	1121.80±161.54
干粉 -18 Dry powder -18	0/5	39.80±11.05*	116.4±23.56*	628.00±158.25*
干粉 -36 Dry powder-36	0/5	42.80±12.4*	133.4±28.41*	832.00±317.07
干粉 -72 Dry powder-72	0/5	43.60±7.64*	174.6±79.81*	956.60±370.99
干粉 -144 Dry powder-144	0/5	80.25±71.79*	128.5±40.22*	844.00±305.04

注：数据为均数 ± 标准差。#$P < 0.05$（与对照组比较）；*$P < 0.05$（与 CCl_4 组比较）。

Data is expressed by mean ± SEM. #$P<0.05$, (compared with the control group); *$P<0.05$ (compared with CCl_4 group).

结论

二月兰籽水煎剂干粉的制备方法保存了药效。

Conclusion

The preparation method of the dry powder of orycho seeds decoction preserves its effection.

大孔树脂分离物的保肝作用

方法

二月兰籽 50kg，10 倍水量分次煎煮两次，1h/ 次，滤液浓缩分次上样（大孔树脂 D101）。

采用水、10%、30%、60% 和 95% 乙醇梯度洗脱。

流分进行动物实验确定有效段。模型药：CCl_4（1.5% 溶于橄榄油），皮下染毒 0.05mL/10g。健康 ICR 小鼠（清洁级）40 只，雄性，体重为 20~22g，随机分为正常组和 CCl_4 模型组，二月兰籽水煎剂组，上样流出物 + 水洗脱物溶液组，10%、30%、60% 和 95% 的乙醇洗脱物溶液组。连续给药 4 天，第 4 天皮下染毒，24h 后取血测 ALT、AST 和 LDH，肝组织做 HE 染色。

Liver-protective activity of macroporous resin separation

Methods

50kg of February orchid seeds were decocted twice with 10 times of water for 1h/time, and the filtrate was concentrated and sampled by times.

Gradient elution with water, 10%, 30%, 60% and 95% ethanol.

The effective fraction was determined by animal experiment. The model drug was CCl_4(1.5% dissolved in olive oil), and the subcutaneous exposure was 0.05mL/10g. Forty healthy ICR mice (clean grade), male, weighing 20~22g, were randomly divided into normal group, CCl_4 model group, orycho seeds decoction group, sample effluent+water eluate solution group, and 10%, 30%, 60% and 95% ethanol eluate solution group. After four days of continuous administration, the rats were exposed subcutaneously on the fourth day. After 24 hours, blood samples were taken to measure ALT, AST and LDH, and liver tissues were stained with HE method.

结果

（1）大孔树脂分离情况

a. 水煎剂出膏率（二月兰出膏率 = 干膏量 / 药材量）为：27×20÷5=108（g/kg）。

b. 各分离物所占百分比如下。

上样流出物：15.42%；水洗脱物：39.79%；10% 乙醇洗脱物：5.81%；30% 乙醇洗脱物：15.54%；60% 乙醇洗脱物：14.92%；95% 乙醇洗脱物：8.52%。

（2）大孔树脂分离物的保肝作用（肝病理）

a. CCl_4 模型组（图 4-7 B）的肝组织出现炎症细胞浸润，以淋巴细胞为主，中央静脉周围充血，肝细胞肿胀、坏死，细胞质 水分含量增多，几乎透明，呈现出气球样变。

b. 二月兰籽水煎剂组和大孔树脂 10% 乙醇洗脱物组（图 4-7 C、图 4-7 E）的 镜下结构和正常组基本一致，未观察到病理改变。

c. 其他洗物组未见改善（图 4-7）。

（3）大孔树脂分离物的保肝作用（肝功酶）

CCl_4 模型组小鼠经皮下注射 1.5% 的 CCl_4 经过 24h 后，小鼠血液中的肝功酶 ALT、AST、LDH 明显升高。在二月兰籽水煎剂和 10% 乙醇洗脱物组中，这些指标水平趋近于正常组，与 CCl_4 模型组相比较，差异有统计学意义（$P<0.05$）。结果见表 4-14。

Results

(1) Macroporous resin separation

a. The extraction rate of decoction (extraction rate orycho seed = dry extract amount/amount of medicinal materials) is : 27×20÷5=108(g/kg).

b. The percentage of each separation is as follows:

Sample effluent: 15.42%; water eluate: 39.79%; 10% ethanol eluate: 5.81%; 30% ethanol eluate: 15.54%; 60% ethanol eluate: 14.92%; 95% ethanol eluate: 8.52%.

(2) Protective effect of macroporous resin separation on liver (pathology)

a. In CCl_4 model group (Fig.4-7 B), the liver tissue showed inflammatory cell infiltration, mainly lymphocytes, congestion around the central vein, swelling and necrosis of liver cells, increased cytoplasm water content, almost transparent and balloon-like changes.

b. The microscopic structures of the water decoction group and the macroporous resin 10% ethanol eluate group (Fig.4-7 C, Fig.4-7 E) were basically the same as those of the normal group, and no pathological changes were observed.

c. No improvement was found in other washing groups（Fig.4-7）.

(3) Protective effect of macroporous resin separation on liver (liver functional enzyme)

In CCl_4 model group, after subcutaneous injection of 1.5% CCl_4 for 24h, the liver function enzymes ALT, AST and LDH in the blood of mice increased significantly. In the water decoction and 10% ethanol eluate group, these indexes were close to the normal group, and the difference was statistically significant compared with CCl_4 model group ($P<0.05$). The results are shown in table 4-14.

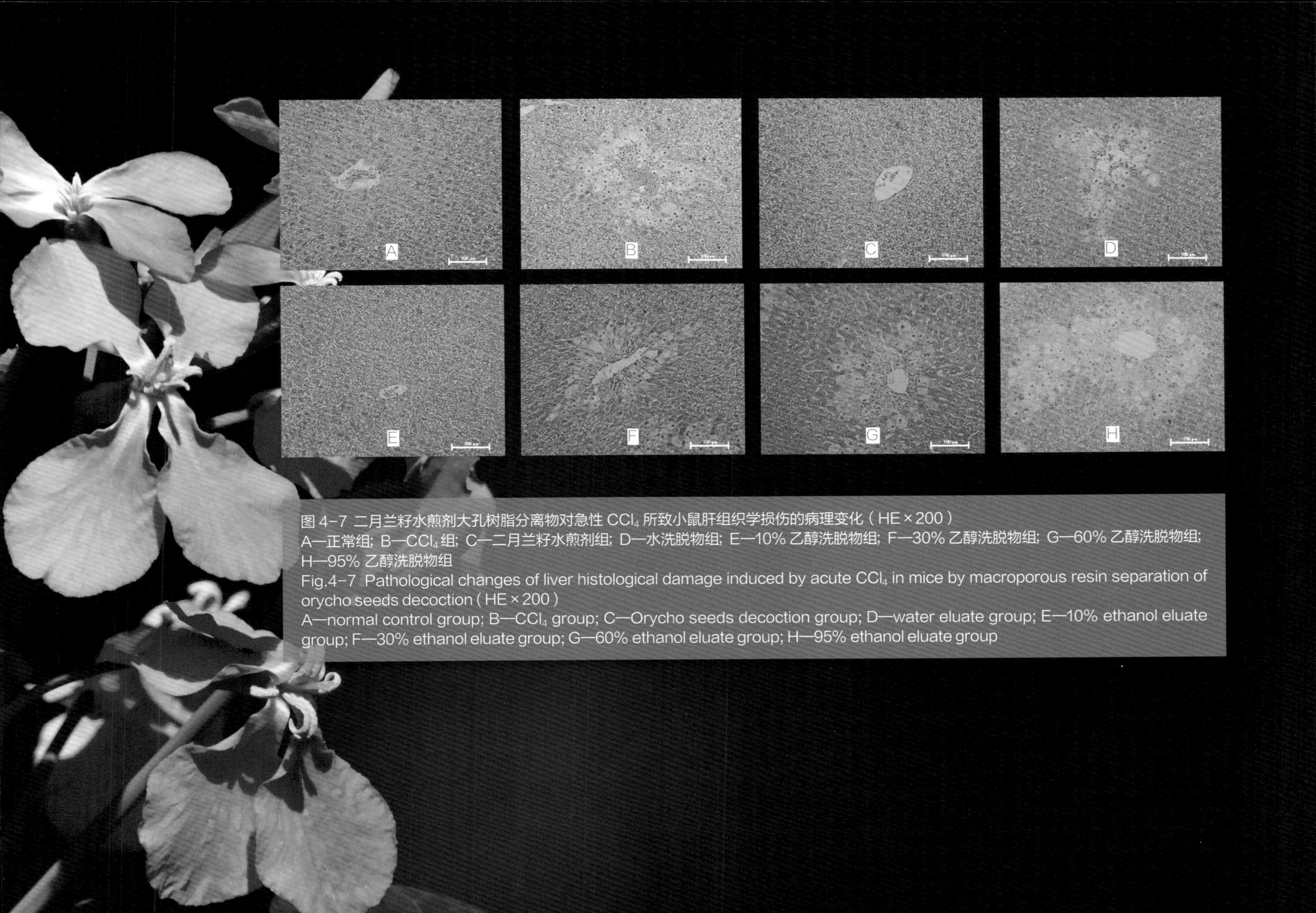

图 4-7 二月兰籽水煎剂大孔树脂分离物对急性 CCl_4 所致小鼠肝组织学损伤的病理变化（HE×200）

A—正常组；B—CCl_4 组；C—二月兰籽水煎剂组；D—水洗脱物组；E—10% 乙醇洗脱物组；F—30% 乙醇洗脱物组；G—60% 乙醇洗脱物组；H—95% 乙醇洗脱物组

Fig.4-7 Pathological changes of liver histological damage induced by acute CCl_4 in mice by macroporous resin separation of orycho seeds decoction (HE×200)

A—normal control group; B—CCl_4 group; C—Orycho seeds decoction group; D—water eluate group; E—10% ethanol eluate group; F—30% ethanol eluate group; G—60% ethanol eluate group; H—95% ethanol eluate group

表 4-14 二月兰籽大孔树脂分离物对 CCl_4 所致小鼠的肝功酶变化的影响

Table 4-14 Effect of macroporous resin extracts from orycho seeds on changes of liver function enzymes in mice Induced by CCl_4

分组 Groups（%）	死亡 Death	ALT（U/L）	AST（U/L）	LDH（U/L）
正常对照组 Normal control	0/5	42.00±7.31	172.00±67.83	1159.60±444.11
CCl_4	0/5	3605.00±3331.45#	1521.67±1311.08#	6216.67±4938.14#
水煎剂 Decoction	0/5	52.14±11.88**	114.57±24.43**	872.14±112.2**
水洗脱物 Water eluate	0/5	3254.29±3471.19	1517.14±1644.06	5490.00±4951.27
乙醇洗脱物 -10% Ethanol eluate-10%	0/5	56.29±8.32**	135.43±29.11**	1086.29±269.17**
乙醇洗脱物 -30% Ethanol eluate-30%	0/5	5204.00±7989.65	2574.00±4617.52	7136.00±11201.53
乙醇洗脱物 -60% Ethanol eluate-60%	0/5	2500.00±1083.44	790.00±522.97	4034.00±3363.83
乙醇洗脱物 -95% Ethanol eluate-95%	0/5	3890.00±3628.87	1414.00±1166.76	6252.00±4937.41

注：数据为均数 ± 标准差。#P<0.05（与正常组相比）；**P<0.01（与模型组相比）；上样流出物 + 水洗脱物组，10%、30%、60% 和 95% 的乙醇洗脱物组的剂量分别为 10.80g/kg、26.65g/kg、4.19g/kg、11.19g/kg、10.74g/kg 和 6.14g/kg，二月兰籽水煎剂组剂量为 72g/kg。

Data is expressed by mean ± SEM. #P<0.05 (compared with the control group); **P<0.01 (compared with CCl_4 group).The effluent+water eluate group, 10%, 30%, 60% and 95% ethanol eluate group were 10.80g/kg, 26.65g/kg, 4.19g/kg, 11.19g/kg, 10.74g/kg and 6.14g/kg, respectively. The dose of the orycho seeds decoction group was 72g/kg.

结论

二月兰籽的保肝有效成分位于 10% 乙醇洗脱物中。

Conclusion

The liver-protecting active components of the orycho seeds are located in 10% ethanol eluate.

反相分离纯化物的保肝作用

方法

将大孔树脂有效段进行反向分离纯化，采用 10%、30%、50%、70% 和 100% 的甲醇梯度洗脱。

流分进行动物实验确定有效段。模型药：CCl_4（1.5% 溶于橄榄油），皮下染毒 0.05mL/10g。健康 ICR 小鼠（清洁级）59只，雄性，体重为20~22g，随机分为正常组和CCl_4模型组，二月兰籽水煎剂组，10%、30%、50%、70% 和 100% 甲醇洗脱物溶液组。连续给药 4 天，第 4 天皮下染毒，24h 后取血测 ALT、AST 和 LDH，肝组织做 HE 染色。

Liver-protective activity of reverse phase separation and purification

Methods

The effective segment of macroporous resin was separated and purified in reverse direction, and eluted with 10%, 30%, 50%, 70% and 100% methanol gradient.

The effective fraction was determined by animal experiment. The model was CCl_4 (1.5% dissolved in olive oil), and the subcutaneous injection was 0.05mL/10g. Fifty-nine healthy ICR mice (clean grade), male, weighing 20~22g, were randomly divided into normal group, CCl_4 model group, orycho seeds decoction group, 10%, 30%, 50%, 70% and 100% methyl alcohol eluent solution group. After four days of continuous administration, the rats were exposed subcutaneously on the fourth day. After 24 hours, blood samples were taken to measure ALT, AST and LDH, and liver tissues were stained with HE.

结果

（1）肝脏病理学结果（图 4-8）

正常对照组（图 4-8 A）肝小叶结构清晰，肝细胞索围绕中央静脉成放射状排列，细胞结构完整清晰，无坏死、变性或炎症细胞浸润。CCl_4 模型组（图 4-8 B）的肝组织可见炎症细胞浸润，以淋巴细胞为主，中央静脉周围可见不同程度的充血，肝细胞可见不同程度的肿胀、坏死，细胞质水分含量增多，几乎透明，呈现出气球样变。二月兰籽水煎剂组和反相分离 10% 甲醇洗脱物组（图 4-8 C、图 4-8 D）的镜下结构和正常组基本一致，未观察到病理改变。

（2）二月兰籽反相高效液相色谱法分离提取物抗小鼠 CCl_4 肝损伤

在二月兰籽水煎剂和反相分离 10% 甲醇洗脱物组中，小鼠血清 ALT、AST、LDH 活性趋近于正常组，与 CCl_4 模型组相比较，差异有统计学意义（$P<0.05$），反相分离 30% 和 50% 甲醇洗脱物组中，小鼠的生化指标与模型组比较差异也有统计学意义。结果见表 4-15。

Results

(1) Liver pathology result (Fig. 4-8)

In the normal control group (Fig.4-8 A), the hepatic lobule structure was clear, the hepatic cords were arranged radially around the central vein, the cellular structure was complete and clear, and there was no necrosis, degeneration or inflammatory cell infiltration. Inflammatory cells infiltrated in the liver tissue of CCl_4 model group (Fig.4-8 B), mainly lymphocytes, with different degrees of congestion around the central vein, swelling and necrosis of liver cells, increased cytoplasm water content, almost transparent and balloon-like changes. The microscopic structures of the orycho seeds decoction group and the reversed-phase separated 10% methanol eluate group (Fig.4-8 C, Fig.4-8 D) were basically the same as those of the normal group, and no pathological changes were observed.

(2) Anti-CCl_4 liver injury in mice by RP-HPLC separation of extracts orycho seeds

The activities of ALT, AST, LDH in serum of mice in the orycho seeds decoction and reversed-phase separated 10% methanol eluate group were close to those in the normal group, and the differences were statistically significant compared with CCl_4 model group ($P<0.05$). the biochemical indexes of mice in the reversed-phase separated 30% and 50% methanol eluate group were also statistically different from those in the model group. The results are shown in table 4-15.

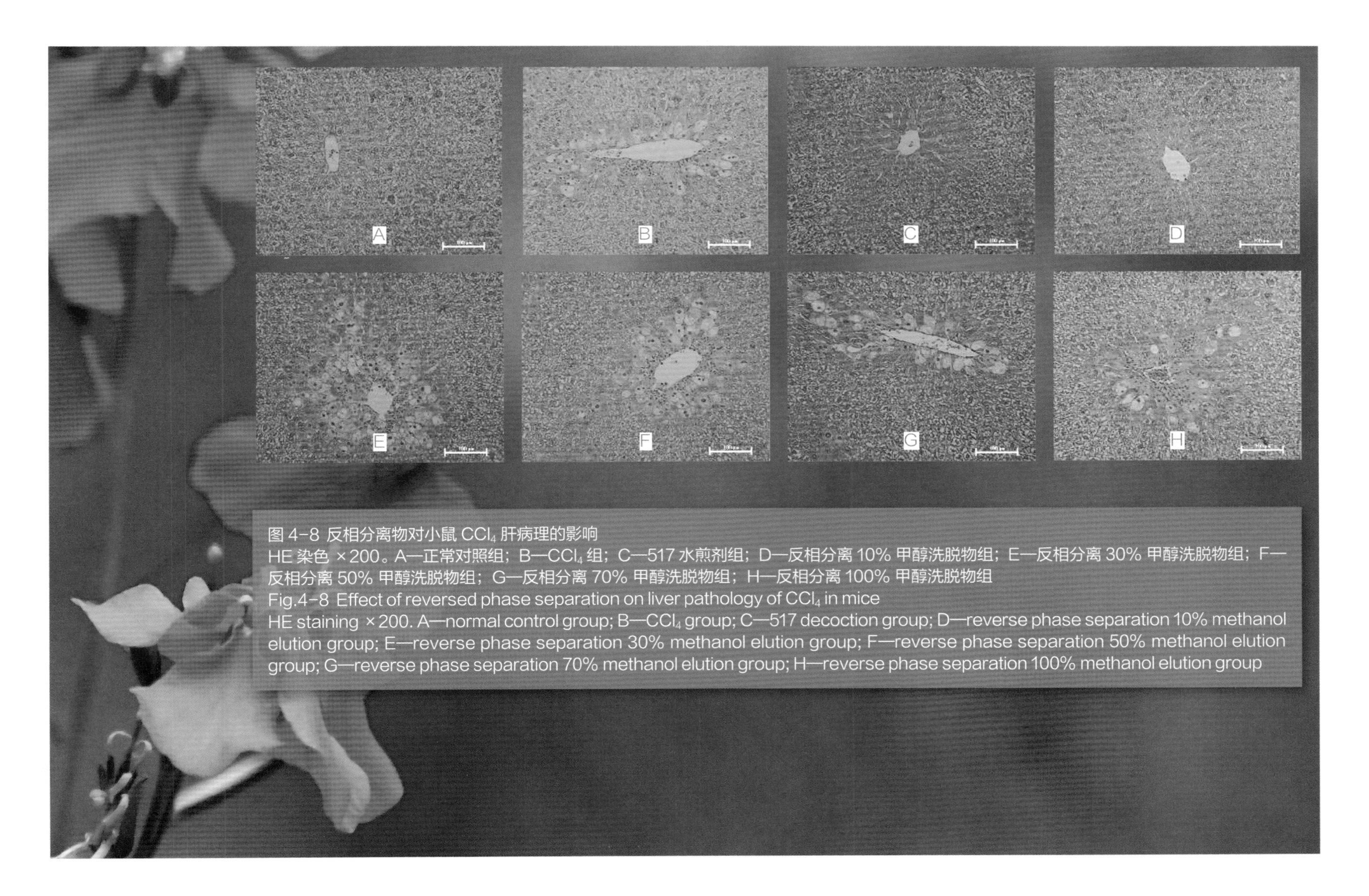

图 4-8 反相分离物对小鼠 CCl_4 肝病理的影响

HE 染色 ×200。A—正常对照组；B—CCl_4 组；C—517 水煎剂组；D—反相分离 10% 甲醇洗脱物组；E—反相分离 30% 甲醇洗脱物组；F—反相分离 50% 甲醇洗脱物组；G—反相分离 70% 甲醇洗脱物组；H—反相分离 100% 甲醇洗脱物组

Fig.4-8 Effect of reversed phase separation on liver pathology of CCl_4 in mice

HE staining ×200. A—normal control group; B—CCl_4 group; C—517 decoction group; D—reverse phase separation 10% methanol elution group; E—reverse phase separation 30% methanol elution group; F—reverse phase separation 50% methanol elution group; G—reverse phase separation 70% methanol elution group; H—reverse phase separation 100% methanol elution group

表 4-15 二月兰籽反相分离物对急性 CCl_4 所致肝损伤小鼠的血清 ALT、AST、LDH 变化

Table 4-15 Effect of reversed phase separation on CCl_4 induced serum liver function enzyme changes

分组 Groups（%）	死亡 Death	ALT（U/L）	AST（U/L）	LDH（U/L）
对照组 Control	0/5	41.80±13.44	102.80±9.12	918.20±217.4
CCl_4	0/12	3034.17±2566.33#	1290±1076.99#	4488.33±3161.75#
水煎剂 -72 Decoction-72	0/7	41.71±11.46**	109.86±19.45*	755.57±123.60**
反相分离 -10% RPC-10%	0/7	41.71±4.75**	98.57±10.26*	711.14±146.37**
反相分离 -30% RPC-30%	0/7	505.71±563.61*	257.14±130.86*	1292.86±443.00*
反相分离 -50% RPC-50%	0/7	562.86±710.93*	318.57±737.60*	1664.29±1036.48*
反相分离 -70% RPC-70%	0/7	1264.29±906.90	687.14±524.97	2830.00±1984.30
反相分离 -100% RPC-100%	0/7	1694.29±612.75	892.86±539.99	3125.71±1657.55

注：数据为均数 ± 标准差。#P<0.05（与对照组相比）；*P<0.05，**P<0.01（与模型组相比）；反相分离10%、30%、50%、70%和100%的剂量分别为1.85g/kg、1.40g/kg、0.68g/kg、0.13g/kg、1.37g/kg，二月兰籽水煎剂组剂量为 72 g/kg。

Data is expressed by mean±SEM. #P<0.05 (compared with the control group); *P<0.05, **P<0.01 (compared with CCl_4 group). The doses of RPC-10%、30%、50%、70%、100% were 1.85g/kg、1.40g/kg、0.68g/kg、0.13g/kg、1.37g/kg, respectively. The dose of the orycho seeds decoction group was 72g/kg.

结论

反相分离物 10% 甲醇段的效果最好。

Conclusion

The reverse phase separation of 10% methanol has the best effect.

凝胶分离物的保肝作用

方法

（1）薄层色谱法（thin layer chromatography，TLC）

在薄层层析法（TLC）中用二氯甲烷和甲醇来配置展开剂，并选取展板上各点里原点距离短的比例来配置硅胶柱的洗脱液，进行二氯烷和甲醇的梯度洗脱。

显色：采用物理检测法。采用 GF254 型硅胶板，用 254nm 波长的紫外光进行检测，将物质相近的进行合并。

（2）凝胶分离法

将反向分离物进行凝胶（Sephadex LH-20）分离，用甲醇洗脱，流速平为 0.2~0.01drop/s，用小青瓶接流分，每 10mL 一瓶，共 20 个。

薄层色谱法检识（图 4-9），合并相同流分。

流分进行动物实验确定有效段。

流分动物实验模型药：CCl_4（1.5% 溶于橄榄油），皮下染毒 0.05mL/10g。健康 ICR 小鼠（清洁级）67 只，雄性，体重为 20~22g，随机分为正常组和 CCl_4 模型组，二月兰籽水煎剂组，凝胶分离段 1、凝胶分离段 2、凝胶分离段 3、凝胶分离段 4、凝胶分离段 5、凝胶分离段 6 和凝胶分离段 7 组。连续给药 4 天，第 4 天皮下染毒，24h 后取血测 ALT、AST 和 LDH，肝组织做 HE 染色。

Liver-protective activity of gel separation

Methods

(1) Thin layer chromatography (TLC)

In thin layer chromatography (TLC), dichloromethane and methanol were used to prepare the developing agent, and the eluent of silica gel column was prepared according to the proportion of short origin distance in each point on the plate, and the gradient elution of dichloromethane and methanol was carried out.

Color development: physical detection method is adopted. GF254 silica gel plate was used for detection by ultraviolet light with wavelength of 254nm, and similar substances were combined.

(2) Gel separation method

Gel separation (Sephadex LH-20) was carried out on the reversed phase separation, and methanol was used for elution, with an average flow rate of about 0.2~0.01drop/s, and small green bottles were used to collect the fractions, one bottle per 10 mL.

Identify by thin layer chromatography (Fig. 4-9), and mix that same fractions.

Animal experiments were carried out to determine the effective fraction.

The model drug of animal experiment was CCl_4 (1.5% dissolved in olive oil), which was exposed subcutaneously at 0.05mL/10g. Sixty- seven healthy ICR mice (clean grade), male, weighing 20~22g, were randomly divided into normal group, CCl_4 model group, orycho seeds decoction group, gel separation stage 1, gel separation stage 2, gel separation stage 3, gel separation stage 4, gel separation stage 5, gel separation stage 6 and gel separation stage 7. After four days of continuous administration, the rats were exposed subcutaneously on the fourth day. After 24 hours, blood samples were taken to measure ALT, AST and LDH, and liver tissues were stained with HE.

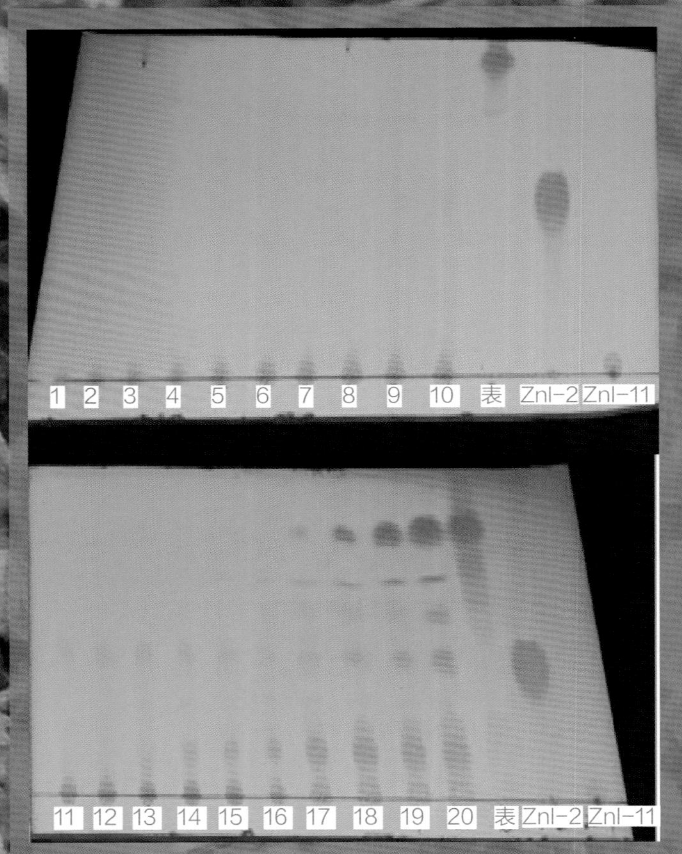

图 4-9 二月兰籽反向分离物的凝胶分离物的薄层色谱法检识图
图中的“表”为表告依春，Znl-2 为二月兰籽水煎剂中的亚精胺 1 型，Znl-11 为二月兰籽水煎剂中的亚精胺 2 型。这些也是我们检测到的二月兰籽中的成分
The " 表 " in the figure is epigoitrin, Znl-2 is spermidine type 1 in the orycho seed decotion, and Znl-11 is spermidine type 2 in the orycho decotion. These are also the components in orycho seed we detected

合并 TCL 相同的流分：

1~4 合并为凝胶分离段 1，5 为凝胶分离段 2，6~7 合并为凝胶分离段 3，8~10 合并为凝胶分离段 4，11~14 合并为凝胶分离段 5，15~16 合并为凝胶分离段 6，17~20 合并为凝胶分离段 7。

各段的比例分别为 0.14%、0.1%、0.29%、0.34%、0.51%、0.48% 和 0.7%。

各段占水煎剂的比例分别为 0.14%、0.1%、0.29%、0.34%、0.51%、0.48% 和 0.7%。

To merge TCL Same Streams:

1~4 are combined into a gel separation section 1, 5 is a gel separation section 2, 6~7 are combined into a gel separation section 3, 8~10 are combined into a gel separation section 4, 11~14 are combined into a gel separation section 5, 15~16 are combined into a gel separation section 6, 17~20 are combined into a gel separation section 7.

The proportion of each section were 0.14%、0.1%、0.29%、0.34%、0.51%、0.48% and 0.7%.

The proportion of each section in decoction were 0.14%、0.1%、0.29%、0.34%、0.51%、0.48% and 0.7%。

结果

（1）肝脏病理学结果

正常对照组（图 4-10 A）肝小叶结构清晰，肝细胞索围绕中央静脉成 放射状排列，细胞结构完整清晰，无坏死、变性或炎症细胞浸润。CCl_4 模型组(图 4-10 B）的肝组织可见炎症细胞浸润，以淋巴细胞为主，中央静脉周围可见不同程度的充血，肝细胞可见不同程度的肿胀、坏死，细胞质水分含量增多，几乎透明，呈现出气球样变。二月兰籽水煎剂组、凝胶分离段 5、6 和 7 组（图 4-10 C、图 4-10 H、图 4-10 I 和图 4-10 J）的镜下结构和正常组基本一致，未观察到病理改变。凝胶分离段 1、2、3 和 4 组的肝组织在光学显微镜下可见中央静脉周围细胞的气球样变，可见细胞质不同程度的肿胀，但病变程度比模型组轻（图 4-10）。

（2）凝胶分离物抗小鼠 CCl_4 肝损伤

CCl_4 模型组小鼠经皮下注射 1.5% 的 CCl_4 经过 24h 后，小鼠血清中的 ALT、AST、LDH 水平升高。在二月兰籽水煎剂、凝胶分离段 5、6 和 7 组，这些指标趋近于正常组，与 CCl_4 模型组相比较，差异有统计学意义（P<0.05）。结果见表 4-16。

Results

(1) Liver pathology result

In the normal control group (Fig.4-10 A), the structure of hepatic lobules is clear, the hepatic cords are arranged radially around the central vein, the cell structure is complete and clear, and there is no necrosis, degeneration or inflammatory cell infiltration. In CCl_4 model group (Fig.4-10 B), inflammatory cells infiltration was observed in the liver tissue, mainly lymphocytes, hyperemia to varying degrees was observed around the central vein, swelling and necrosis to varying degrees were observed in the liver cells, and the water content in the cytoplasm increased, almost transparent, presenting balloon-like changes. The microscopic structures of the orycho seeds decoction group and gel separation sections 5, 6 and 7 groups (Fig.4-10 C, Fig.4-10 H, Fig.4-10 I and Fig.4-10 J) were basically the same as those of the normal group, and no pathological changes were observed. Balloon-like changes of cells around the central vein can be seen in the liver tissue of gel separation sections 1, 2, 3 and 4 groups under the optical microscope, and swelling of cytoplasm can be seen to varying degrees, but the lesion degree is lighter than that of the model group (Fig. 4-10).

(2) Anti-CCl_4-induced liver injury in mice by gel separation

After 24 h of subcutaneous injection of 1.5% CCl_4, the levels of ALT, AST and LDH in mice serum increased. In orycho seeds decoction and gel separation sections 5, 6, 7 groups, these indexes approach to the normal group, and the difference is statistically significant (P<0.05) when compared with the CCl_4 model group. The results are shown in Table 4-16.

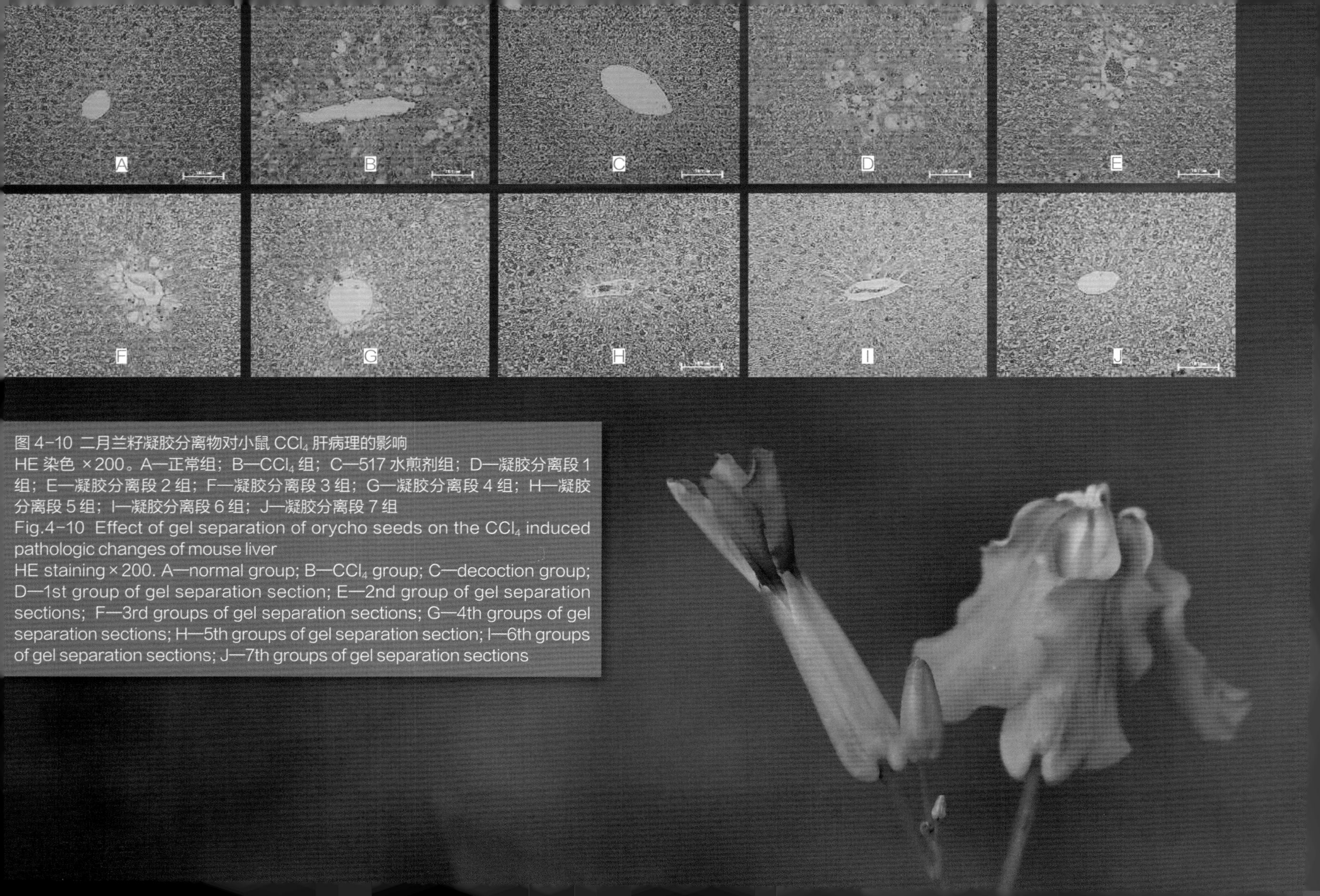

图 4-10 二月兰籽凝胶分离物对小鼠 CCl_4 肝病理的影响

HE 染色 ×200。A—正常组；B—CCl_4 组；C—517 水煎剂组；D—凝胶分离段 1 组；E—凝胶分离段 2 组；F—凝胶分离段 3 组；G—凝胶分离段 4 组；H—凝胶分离段 5 组；I—凝胶分离段 6 组；J—凝胶分离段 7 组

Fig.4-10 Effect of gel separation of orycho seeds on the CCl_4 induced pathologic changes of mouse liver

HE staining × 200. A—normal group; B—CCl_4 group; C—decoction group; D—1st group of gel separation section; E—2nd group of gel separation sections; F—3rd groups of gel separation sections; G—4th groups of gel separation sections; H—5th groups of gel separation section; I—6th groups of gel separation sections; J—7th groups of gel separation sections

表 4-16 凝胶分离物对血清肝功酶的影响
Table 4-16 Effect of gel separation sections on CCl_4 induced serum liver function enzyme changes

分组 Groups（g/kg）	死亡 Death	ALT（U/L）	AST（U/L）	LDH（U/L）
正常对照组 Normal control	0/5	35.40±3.36	35.40±3.36	870.00±216.12
CCl_4	0/6	363.33±320.29#	363.33±320.29#	1516.67±357.64##
水煎剂 -72 Decoction-72	0/7	55.00±23.40**	55.00±23.40**	958.14±273.13**
凝胶分离段 -1 Section-1	0/7	550.57±538.03	550.57±538.03	1053.86±225.34
凝胶分离段 -2 Section-2	0/7	397.86±592.07	397.86±592.07	1086.00±528.31
凝胶分离段 -3 Section-3	0/7	486.43±638.67	486.43±638.67	1036.00±526.27
凝胶分离段 -4 Section-4	0/7	531.00±517.29	531.00±517.29	1224.43±531.41
凝胶分离段 -5 Section-5	0/7	45.86±14.77**	45.86±14.77**	591.57±81.46**
凝胶分离段 -6 Section-6	0/7	39.00±6.56**	39.00±6.56**	590.14±249.32**
凝胶分离 -7 Section-7	0/7	41.43±13.91**	41.43±13.91**	529.86±117.75**

注：数据为均数 ± 标准差。#P<0.05，##P<0.01（与正常组相比）；**P<0.01（与模型组相比）；凝胶分离段1、2、3、4、5、6和7组的剂量分别为0.11g/kg、0.08g/kg、0.21g/kg、0.25g/kg、0.37g/kg、0.35g/kg和0.51g/kg，二月兰籽水煎剂组剂量为72g/kg。
Data is expressed by mean ± SEM. #P<0.05 (compared with the control group); ##P<0.01 (compared with CCl_4 group).
The doses of gel separation section 1, 2, 3, 4, 5, 6 and 7 were 0.11g/kg, 0.08g/kg, 0.21g/kg, 0.25g/kg, 0.37g/kg, 0.35g/kg and 0.51g/kg, respectively, and the dose of the orycho seeds decoction group was 72g/kg.

结论

凝胶分离段 5、6 和 7 组有保肝活性。

Conclusion

Section 5, 6 and 7 of the gel separation have hepatoprotective activity.

半制备型高效液相色谱分离物的保肝作用

方法

配置洗脱液：10%~15% 乙腈（1% 甲酸）。

条件：柱温 25℃，紫外检测波长 210nm。

进样：进样量 200μL，流速 8mL/min。

采集：除去第一个溶剂峰，分别回收不同流分。

旋蒸：分别将液体中的溶剂旋蒸掉。

将流分进行动物实验确定有效段。

流分动物实验造模：CCl_4（1.5% 溶于橄榄油），皮下注射 0.05mL/10g。健康 ICR 小鼠（清洁级）64 只，雄性，体重为 20~22g，随机分为正常组和 CCl_4 模型组，新化合物单独组、新化合物组、联苯双酯组。连续给药 4 天，第 4 天皮下染毒，24h 后取血测 ALT、AST 和 LDH，肝组织做 HE 染色。

动物实验的数据处理：SPSS 18.0 进行统计分析，用均 数 ± 标准差，即$\bar{x} \pm S$ 表示，用单因素方差分析进行处理。

Liver protective activity of semi-preparative high performance liquid chromatography separation

Methods

Prepare eluent: 10%~15% acetonitrile (1% formic acid).

Conditions: column temperature 25℃, ultraviolet detection wavelength 210nm.

Sample injection: the injection volume is 200μL, and the flow rate is 8mL/min.

Collection: remove the first solvent peak and recover different fractions respectively.

Rotary evaporation: the solvents in the liquid are respectively evaporated by rotary evaporation.

The effective segment was determined by animal experiment.

Animal model of fraction experiment: CCl_4 (1.5% dissolved in olive oil), subcutaneous injection 0.05mL/10g. Sixty-four healthy ICR mice (clean grade), male, weighing 20~22g, were randomly divided into normal group, CCl_4 model group, new compound alone group and new compound group. After four days of continuous administration, the rats were exposed subcutaneously on the fourth day. After 24 hours, blood samples were taken to measure ALT, AST and LDH, and liver tissues were stained with HE method.

Data processing of animal experiment: SPSS 18.0 was used for statistical analysis, which was expressed by the mean ± standard deviation, that is, $\bar{x} \pm S$, and processed by one-way analysis of variance.

结果

用半制备型高效液相对该段进行物质单体分离，得到了表告依春、*N*- 苯甲酰基 - 表告依春、新化合物、*N*-2- 羟基 -3- 丁烯基 - 苯甲酰胺、2- 氨基 -5- 羟基苯甲酸和 α -[(2- 乙酰羧基) 氨基]- 苯乙酸六个物质。

（1）肝脏病理学结果

正常对照组（图 4-11 A）肝小叶结构清晰，肝细胞索围绕中央静脉成放射状排列，细胞结构完整清晰，无坏死、变性或炎症细胞浸润。CCl_4 模型组（图 4-11 C）的肝组织可见炎症细胞浸润，以淋巴细胞为主，中央静脉周围可见不同程度的充血，肝细胞可见不同程度的肿胀、坏死，细胞质水分含量增多，几乎透明，呈现出气球样变。新化合物单独组、二月兰籽水煎剂组和新化合物实验组（图 4-11 B、图 4-11 D、图 4-11 E）的镜下结构和正常组基本一致（图 4-11 ）。

（2）新化合物抗小鼠 CCl_4 肝损伤

在二月兰籽水煎剂、新化合物组中，小鼠血清中的肝功酶 ALT、AST、LDH 水平趋近于正常组，与 CCl_4 模型组相比较，差异有统计学意义（$P<0.05$）。结果见表 4-17。

Results

Six substances including epigoitrin, *N*-benzoyl-epigoitrin, a nonel compound, *N*-2-hydroxy-3-butenyl-benzamide, 2-amino - 5-hydroxybenzoic acid and α -[(2- acetocarboxy) amino]- phenylacetic acid were obtained by separating substances and monomers with semi-preparative high-performance liquid.

(1) Liver pathology result

In the normal control group (Fig. 4-11 A), the hepatic lobule structure was clear, the hepatic cords were arranged radially around the central vein, the cellular structure was complete and clear, and there was no necrosis, degeneration or inflammatory cell infiltration. Inflammatory cells infiltrated in the liver tissue of CCl_4 model group (Fig.4-11 C), mainly lymphocytes, with different degrees of congestion around the central vein, swelling and necrosis of liver cells, increased cytoplasm water content, almost transparent and balloon-like changes. The microscopic structures of new compound alone group, the orycho seeds decoction group and the new compound experimental group (Fig.4-11 B, Fig.4-11 D and Fig.4-11 E) are basically the same as those of the normal group (Fig.4-11).

(2) Anti-CCl_4 induced liver injury by the new compound in mice

The levels of liver function enzymes ALT, AST and LDH in the serum of mice in the dosage groups of the orycho seeds decoction and the new compound were close to those in the normal group, and the difference was statistically significant compared with the CCl_4 model group ($P<0.05$). The results are shown in Table 4-17.

图 4-11 新化合物对小鼠 CCl_4 肝病理的影响

HE 染色 ×200。A—正常组；B—新化合物单独组；C—CCl_4 组；D—二月兰水煎剂组；E—新化合物实验组；F—联苯双酯组

Fig.4-11 Effect of the new compound on liver pathology of CCl_4 in mice

HE staining × 200. A—normal group; B—separate group of the new compound; C—CCl_4 group; D—orycho seeds decoction group; E—experimental group of the new compound; F—bifendate group

表 4-17 新化合物对血清肝功酶的影响

Table 4-17 Effect of the new compound on CCl_4 induced serum liver function enzyme changes

分组 Groups（g/kg）	死亡 Death	ALT（U/L）	AST（U/L）	LDH（U/L）
正常对照组 Normal control	0/4	56.75±32.44	127.75±46.98	1089.00±218.03
新化合物单独组 -0.05 New compound separate-0.05	0/4	42.50±5.57	115.50±13.4	1023.50±183.27
CCl_4	0/9	1484.11±599.81##	55.00±23.40##	958.14±273.13##
水煎剂 -72 Decoction-72	0/7	58.86±14.89**	133.14±24.9*	998.29±283.21**
新化合物实验组 -0.05 New compound experimental-0.05	0/7	197.67±214.31**	174.00±49.51*	998.33±317.58**
联苯双酯 -0.05 Bifendate-0.05	0/7	1129.46±628.13	1047.31±383.14	2234.23±777.38*

注：数据为均数 ± 标准差。##$P<0.01$（与正常组相比）；*$P<0.05$，**$P<0.01$（与模型组相比）；二月兰籽水煎剂组剂量为 72g/kg。

Data is expressed by mean ± SEM. ##$P<0.01$ (compared with the control group); *$P<0.05$, **$P<0.01$ (compared with CCl_4 group). The doses of the orycho seeds decoction group was 72g/kg.

结论

新化合物有保肝活性。

Conclusion

The new compound has hepatoprotective activity.

保肝有效成分结构测定

核磁共振法

（1）仪器

核磁共振仪，NMR，AV Ⅲ 600，Bruker 核磁共振仪。

（2）方法

将每种单体物质取 10mg，溶解于 5mL 的 DMSO 中，混匀，用 NMR 测定。

（3）结果

一种新的苷类化合物(表4-18，图4-12~图4-15)。

Structure identification of effective components

Nuclear magnetic resonance method

(1) Instrument

NMR instrument, NMR, AV Ⅲ 600, Bruker NMR instrument.

(2) Methods

Each monomer substance was taken as 10mg, dissolved in 5mL of DMSO, mixed, and determined by NMR.

(3) Result

A new nucleoside compound (Table 4-18, Fig. 4-12~Fig. 4-15).

表 4-18 新化合物二月兰苷的核磁共振数据

Table 4-18 Nuclear magnetic resonance data of the new compound—orychonin

Position	1 (Znl-z-3)	
	δ_C, type	δ_H(J in Hz)
1	156.0, C	
2	149.0, C	
3	140.0, CH	8.35, s
4		
5	152.2, CH	8.14, s
6	119.3, C	
7		
8	156.3, C	
9	117.3, CH	7.43, d (3.0)
10	113.0	
11	149.7, C	
12	117.7, CH	6.88, d (8.4)
13	125.2, CH	7.25, dd (9.0, 3.0)
14	171.5, C	
1’	87.9, CH	5.87, dd (6.6)
2’	73.4, CH	4.61, t (6.0, 5.4)
3’	70.6, CH	4.14, dd (9.0, 9.0)
4’	85.9, CH	3.96, d (3.6)
5’	61.6, CH_2	3.55, dd (6.0, 5.4)
1”	102.0, CH	4.70, d (7.8)
2”	73.2, CH	3.20, m
3”	76.4, CH	3.25, m
4”	69.6, CH	3.15, m
5”	77.0, CH	3.27, m
6”	60.65, CH_2	3.47, dd (6.0, 5.4)
2-NH		7.40, brs

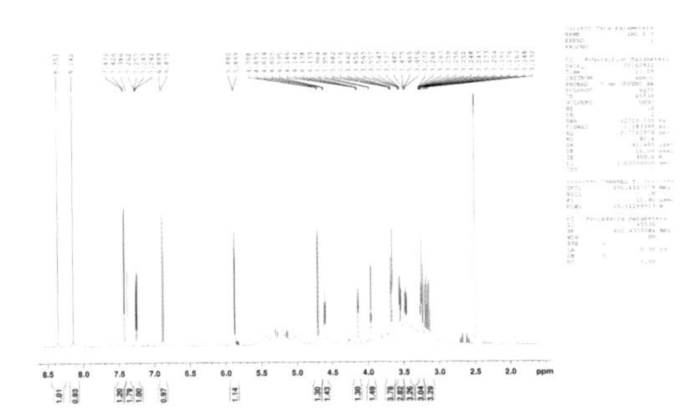

图 4-12 新化合物二月兰苷 1H-NMR
Fig.4-12 1H-NMR of the new compound—orychonin

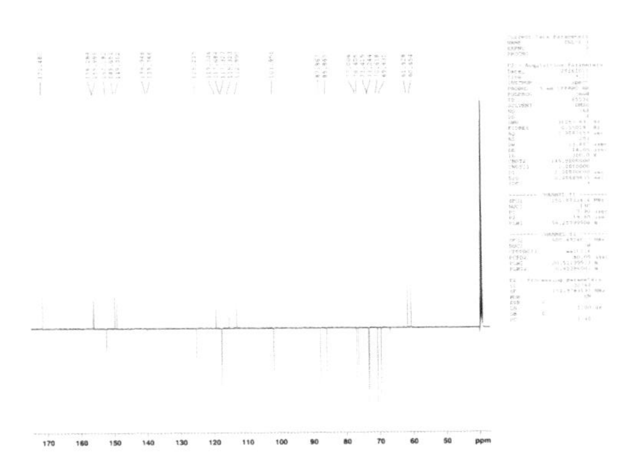

图 4-13 新化合物二月兰苷 13C-NMR
Fig.4-13 13C-NMR of the new compound—orychonin

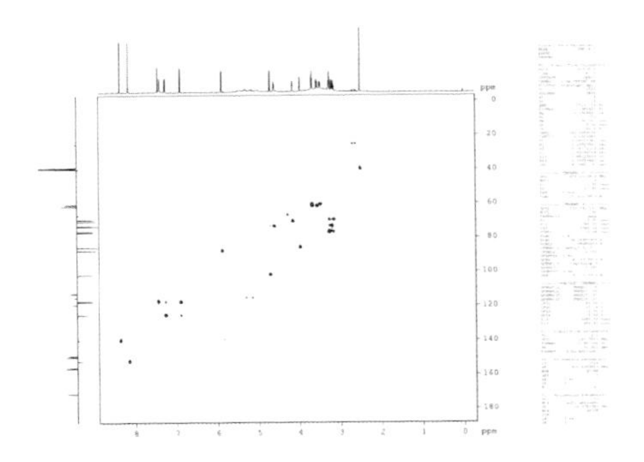

图 4-14 新化合物二月兰苷 HSQC
Fig.4-14 HSQC of the new compound—orychonin

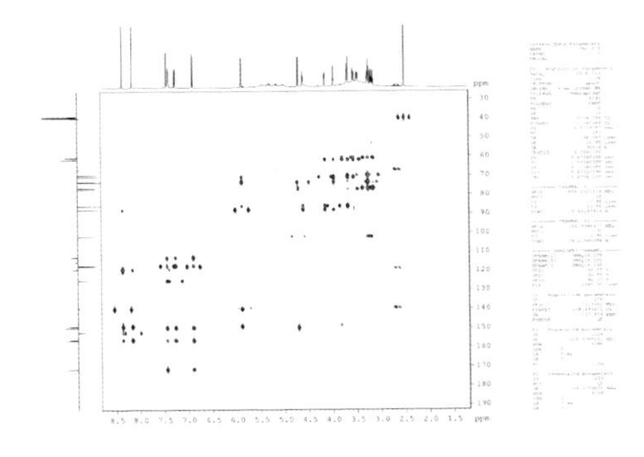

图 4-15 新化合物二月兰苷 HMBC
Fig.4-15 HMBC of the new compound—orychonin

保肝有效成分结构测定

单晶衍射法

（1）仪器

a. 单晶衍射仪 MicroMax003 / XtaLAB Synergy。

b. 分析软件 olex2。

（2）结果

有效成分的单晶衍射图见图4-16。

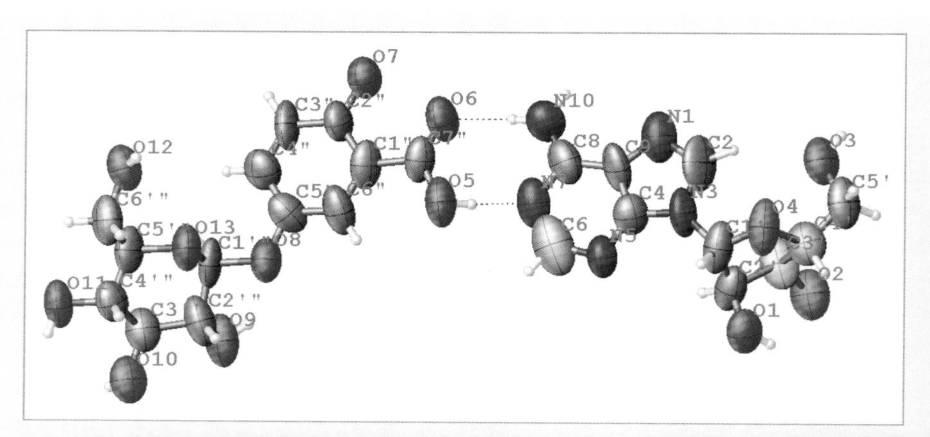

图 4-16 有效成分的单晶衍射图

Fig.4-16 Single crystal diffraction pattern of effective component

Structure identification of effective components

Single crystal diffraction method

(1) Instrument

a. Single crystal diffrac tometer MicroMax003/XtaLAB Synergy.

b. Analysis software olex2.

(2) Result

Single crystal diffraction pattern of effective component see Fig.4-16.

保肝有效成分二月兰苷含量测定

方法

二月兰籽水煎剂新化合物单体含量测定

a. 标准品：取新化合物标准品 3.2mg，用 50% 甲醇溶解定容至 10mL 浓度为 0.32mg/mL。

b. 样品：取水煎剂干粉 18mg，用 30% 甲醇溶解定容至 10mL，浓度为 1.8mg/mL；取 10% 大孔树脂乙醇洗脱物干粉 14.8mg，分别用 50% 甲醇溶解定容至 10mL，浓度为 1.48mg/mL；取 10% 反相分离甲醇洗脱物干粉 17.7mg，分别用 50% 甲醇溶解定容至 10mL，浓度为 1.77mg/mL。

采用超高压液相色谱法进行测定。

Concentration of effective component — orychonin

Methods

Determination of monomer content of new compounds in orycho seeds decoction

a. Standard: 3.2mg of the new compound (orychonin)standard was dissolved in 50% methanol to a 10 mL concentration of 0.32mg/mL.

b. Sample: 18mg dry powder of orycho seeds decoction was dissolved in 30% methanol to a constant volume of 10mL, with the concentration of 1.8mg/mL. 14.8mg of 10% dry powder eluted with macroporous resin ethanol was dissolved in 50% methanol to a constant volume of 10mL, with the concentration of 1.48mg/mL. Each of 17.7mg of the 10% dry powder inversed phase methanol eluate was dissolved in 50% methanol to a volume of 10mL at 1.77mg/mL.

The ultra high pressure liquid chromatography (UPLC) was used.

结果

在257nm 下，新化合物二月兰苷的吸收峰达到最大，保留时间(retention time, TR) 约为 3.4min。

在进样量分别为 0.2 μ L、0.4 μ L、0.6 μ L、0.8 μ L、1.0 μ L、1.2 μ L、1.4 μ L、1.6 μ L 的时候取吸收峰面积，制作标准曲线，标准曲线结果见表 4-19 和图 4-17。 读取水煎剂、10% 大孔树脂乙醇洗脱物和 10% 反相分离甲醇洗脱物样品中新化合物二月兰苷的峰值，分别为 38578、168274 和 809088，得到新化合物二月兰苷在水煎剂、10% 大孔树脂乙醇洗脱物和 10% 反相分离甲醇洗脱物中的含量分别为 0.0529% 、 0.8164% 和 3.7898%，二月兰籽水煎剂的出膏率为 108g/kg，则通过计算得到新化合物二月兰苷在二月兰籽药材中的含量为 0.0057%，见图 4-18~ 图 4-21。新化合物二月兰苷见图 4-22。新化合物和水煎剂的 TLC 检识结果见图 4-23。

Results

At 257nm, the absorption peak of the new compound—orychonin reaches its maximum, and the retention time (TR) is about 3.4min.

The absorption peak areas were taken at the injection volumes of 0.2μL, 0.4μL, 0.6μL, 0.8μL, 1.0μL, 1.2μL, 1.4μL and 1.6μL, respectively, to prepare the standard curves. The results of the standard curves are shown in Table 4-19 and Fig. 4-17 The peak values of the new compound—orychonin in the orycho seeds decoction, the 10% macroporous resin ethanol eluate and the 10% reversed-phase separated methanol eluate samples were read as 38578, 168274 and 809088, respectively, and the contents of the new compound—orychonin in the orycho seeds decoction, the 10% macroporous resin ethanol eluate and the 10% reversed- phase separated methanol eluate were obtained as 0.0529%, 0.8164% and 3.7898%, respectively. the extraction rate of the orycho seeds decoction was 108g/kg, and the new compound orychonin in the orycho seed is 0.0057%. See Fig.4-18~Fig.4-21. Novel compound orychonin see Fig.4-22. Identification results of the new compound and decoctions by TLC see Fig.4-23.

表 4-19 标准曲线数据结果

Table 4-19 Standard curve data results

进样量 Sample size (μL)	换算后的进样量 Sample size after conversion ($\times10^{-4}$ mg)	峰面积 Peak area (mV)
0.2	0.64	206855
0.4	1.28	523775
0.6	1.92	843347
0.8	2.56	1073107
1.0	3.2	1237990
1.2	3.84	1541310
1.4	4.48	1753583
1.6	5.12	1988685

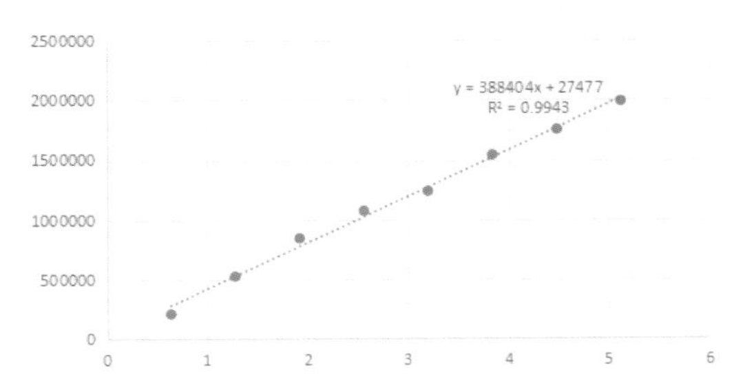

图 4-17 新化合物标准曲线 [X 轴：换算后的进样量（$\times10^{-4}$ mg），Y 轴：响应值（mV）]

Fig.4-17 Scurve of the new compound [X axis: converted injection volume ($\times10^{-4}$ mg), Y axis: response value (mV)]

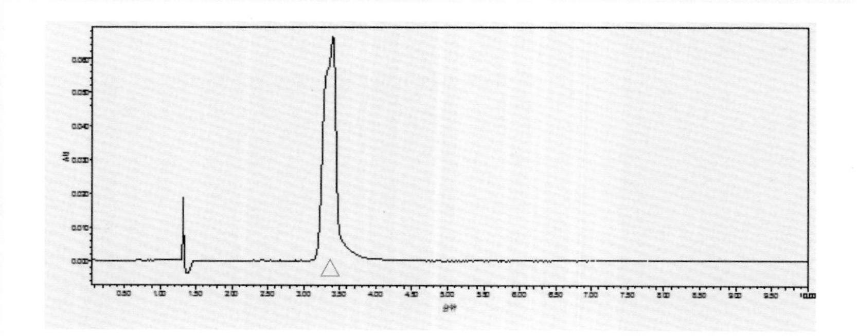

图 4-18 新化合物标准品的紫外吸收图谱（如 Δ 所示）
Fig.4-18 UV absorption spectrum of the new compound standard (as shown in Δ)

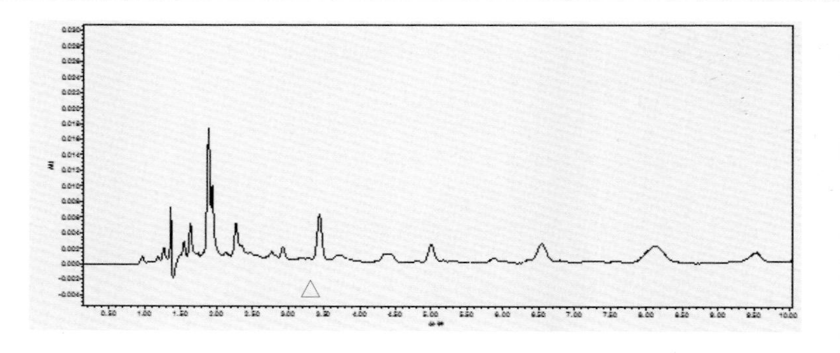

图 4-19 新化合物在水煎剂样品中的紫外吸收图谱（如 Δ 所示）
Fig.4-19 UV absorption spectra of the new compound in decoction samples (as shown in Δ)

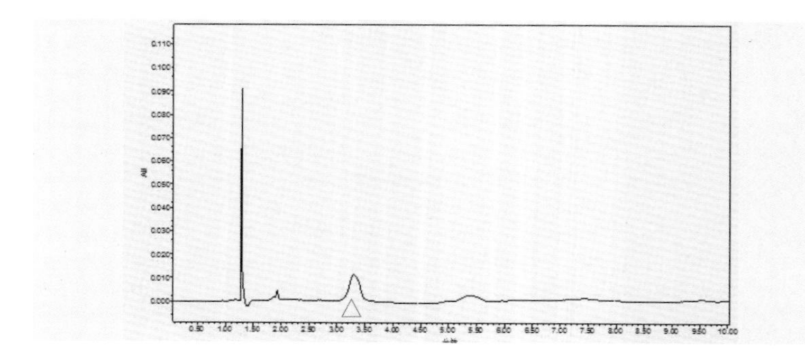

图 4-20 新化合物在 10% 大孔树脂乙醇洗脱物样品中的紫外吸收图谱（如 Δ 所示）
Fig.4-20 Ultraviolet absorption spectra of the new compound in 10% macroporous resin ethanol eluate samples (as shown in Δ)

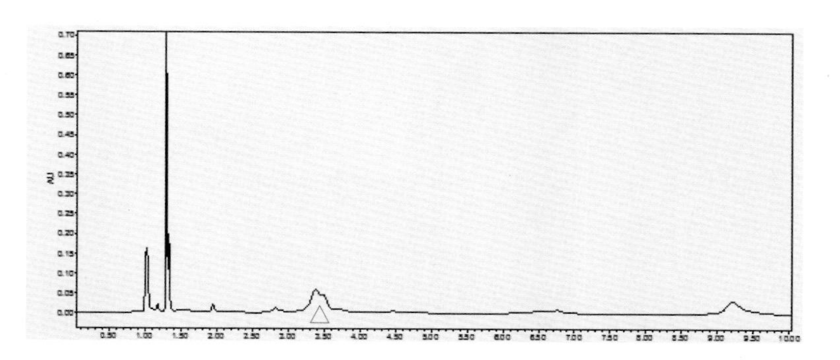

图 4-21 新化合物在 10% 反相分离甲醇洗脱物样品中的紫外吸收图谱（如 Δ 所示）
Fig.4-21 UV absorption spectra of the new compounds in a 10% reversed-phase methanol eluate (as shown in Δ)

图 4-22 新化合物——二月兰苷
Fig.4-22 Novel compound——Orychonin

图 4-23 新化合物和水煎剂的 TLC 检识结果
Fig.4-23 Identification results of the new compound and decoctions by TLC

后记

——二月兰研究历程

好奇心是什么？我们的团队对二月兰的认识和研究过程就回答了这个问题。

时间过得真快，最初与二月兰相遇是十七年前的春天，当年孩子参加学校组织的香山春游后带回来一张照片，问照片上的小花是什么花。我爱人也说在公园里见过这种花，并问我能否对这种花进行研究，也许会有什么新发现。我虽然也不认识这种花叫什么名字，但是对这种花也就是二月兰的研究由此开始。

Postscript

—— A brief Overview of our research on orycho

What is curiosity? Our team story of orycho exploration can probably answer this question.

How time flies! I first met orycho in the spring 17 years ago. When my son took a picture after attending school spring outing to the Fragrant Hill and asked what the little flower in the picture was. My wife also said that she had seen this flower in a park, and asked me if I could study this flower, maybe could find something new. Although I didn't know the name of this flower, the research on this flower, that is, orycho, began.

图 1 我们的部分队员在校园的二月兰小小植物园
Fig.1 Some of the team members were in our orycho garden in campus

图 2 五名研究生的毕业论文
Fig.2 Graduation Thesis of Five Postgraduates

到目前为止，团队一起从不认识二月兰到自己在校园播种小小药用植物园（图 1）、从有用性到安全性研究、一直到寻找其中的有效成分单体、从独立研究和广泛合作。已经有了五名研究生以研究二月兰的成果为毕业论文（图 2）并成功参加了研究生校园科技竞赛（图 3）。漫长的时间和蚂蚁啃骨头的精神磨炼了团队的性格，也使得我们从单一的毒物研究领域开拓到了植物解毒剂领域。在此过程中，大家还一起努力争取到了国家中医药局中药配伍减毒重点研究室，这是北京大学公共卫生学院第一个部委级的研究中心。二月兰的研究经验也为我开设选修课《天然产物开发与安全性评价》打下了良好的基础。这本书也是这个课程的辅助教材。人类的卫生离不开天然植物来源的食品和药物以及化妆品。本书收集了我们的二月兰阶段性研究成果。

Up to now, our team has experienced a shift from knowing nothing about orycho to growing orycho on campus (Fig.1), from finding its health usefulness to its effective component, from independent research activity to extensive cooperating with other teams. Five postgraduate students majored in the research of orycho as their graduation theses (Fig. 2) and won the campus scientific competition (Fig. 3). Our personal character has been moulded by the unremitting efforts in carrying out the research just like an ant gnawing at bones. Our research has extended from simple toxicity identification to the field of searching poison antidote. And we are very honored to be part of the State Key Laboratory of Herbs Detoxification Combination of the State Administration of Traditional Chinese Medicine, which is the first research center of Peking University School of Public Health at ministerial level. The research experience of orycho has also become the lecture content of my campus course "Development and Safety Evaluation of Natural Products". This book is also an auxiliary teaching material for this course. Natural plants are the indispensable treasure of foods, medicines and cosmetics for human health. This book collects our research results of the beautiful herb orycho.

荣誉证书

张宝旭老师：

您被评为北京大学第十七届"挑战杯"——五四青年科学奖竞赛"优秀指导老师"。

特发此证，以资鼓励。

共青团北京大学委员会
北京大学"挑战杯"科技工程办公室
二〇〇九年五月

图3 研究生（下）参加校园挑战杯竞赛的导师鼓励奖（上）
Fig.3 Tutor Award (above)after postgraduate students (below) win the campus competition